Crystal Growth of Multifunctional Borates and Related Materials

Crystal Growth of Multifunctional Borates and Related Materials

Special Issue Editor

Nikolay I Leonyuk

MDPI • Basel • Beijing • Wuhan • Barcelona • Belgrade

MDPI

Special Issue Editor
Nikolay I Leonyuk
M. V. Lomonosov Moscow State University
Russia

Editorial Office
MDPI
St. Alban-Anlage 66
4052 Basel, Switzerland

This is a reprint of articles from the Special Issue published online in the open access journal *Crystals* (ISSN 2073-4352) in 2019 (available at: https://www.mdpi.com/journal/crystals/special_issues/ multifunctional_borates)

For citation purposes, cite each article independently as indicated on the article page online and as indicated below:

LastName, A.A.; LastName, B.B.; LastName, C.C. Article Title. *Journal Name* **Year**, *Article Number, Page Range.*

ISBN 978-3-03897-838-1 (Pbk)
ISBN 978-3-03897-839-8 (PDF)

Contents

About the Special Issue Editor

Nikolay I. Leonyuk graduated from Lomonosov Moscow State University (LMSU) in 1969. He obtained a Ph.D. degree in Crystallography and Crystal Physics, awarded by the Geological Faculty of LMSU, in 1972. He also obtained a Doctor of Science degree in Inorganic Chemistry at the Faculty of Chemistry, LMSU, in 1985. Currently, he is Professor and Scientific Supervisor of Crystallography and Crystal Growth Laboratory of LMSU, awarded Distinguished Professor of Moscow University (2017). His research primarily concerns crystal growth, crystallography, crystal chemistry, and crystal physics of minerals and artificial materials. He had designed several long-term courses of lectures on the crystal growth and characterization and teaches for bachelor, Master and PhD students—crystallographers and gemologists.

crystals

MDPI

Editorial

Crystal Growth of Multifunctional Borates and Related Materials

Nikolay I Leonyuk

Department of Crystallography and Crystal Chemistry, Moscow State University, 119992 Moscow, Russia;
leon@geol.msu.ru

Received: 15 March 2019; Accepted: 19 March 2019; Published: 21 March 2019

Keywords: crystal growth; crystallography; crystal chemistry; borates; multifunctional materials

Crystalline materials play an important role in modern physics and electronics. Therefore, the demand for crystals with functional properties is increasing strongly, due to the technical advance in different fields: telecommunications, computer devices, lasers, semiconductors, sensor technologies, etc. At the first stage, natural minerals (e.g., quartz) were widely used as piezoelectric and optical material. Later on, after the creation of the first laser, interactions between lasers and materials have been investigated: radiation at the double the frequency of a ruby laser was observed as the fundamental light passing through a quartz crystal [1]. This phenomenon became a substantial contribution to the field of quantum electronics and nonlinear optics. However, natural single crystals usually have insufficient purity, size, occurrence, and homogeneity, or do not even exist in nature. That is why the material scientists began to develop important basic materials with the desirable properties. As an example, at the beginning of the 1960s, this resulted in Czochralski growth of $Y_3Al_5O_{12}$ crystals, referred to as YAG, which is the progenitor of the large group of synthetic materials belonging to the structural type of natural garnet family $A_3B_2(SiO_4)_3$ [2]. Owing to a reasonable growth technology, these crystals and their numerous derivatives including transparent nano-ceramics are dominating the elemental base for solid-state laser engineering and various practical applications.

In the meantime, natural and even highly technological synthetic crystals have reached the limit of their potential for fast-progressing science and engineering. The creation of new crystals with predictable structures and, therefore, desirable physical characteristics is restrained by the theoretical, methodological, and technical problems connected with their crystallization from multicomponent systems. Among them, more than 1000 representatives of the anhydrous borate family are listed in the Inorganic Crystal Structure Database [3]. These compounds are characterized by the great variety in their crystal structures, caused in the linkage of planar BO_3–triangles and BO_4–tetrahedra as fundamental structural units. This also leads to glass formation in viscous borate-based melts. Therefore, investigations of "conditions–composition–structure-properties" relationships can help to develop the technology of single crystal components for high performance electronic and optical devices for industrial, medical and entertainment applications. These research works have quickly opened a new field of materials science.

Most of the borate materials attract considerable attention owing to their remarkable characteristics and potential applications. For instance, they demonstrate nonlinear optical and piezoelectric effects (CsB_3O_5, LiB_3O_5, $CsLiB_6O_{10}$, $KBe_2BO_3F_2$, $Sr_2Be_2BO_7$, $K_2Al_2B_2O_7$, $Ca_4GdO(BO_3)_3$, β-BaB_2O_4, $R_2CaB_{10}O_{19}$, $RM_3(BO_3)_4$, where R – rare-earth elements; M – Al, Cr, Ga, Fe, Sc) [4–6], etc., luminescent (RBO_3) [7–9] and magneto-electrical properties ($RFe_3(BO_3)_4$, $RCr_3(BO_3)_4$, $HoAl_3(BO_3)_4$, $TbAl_3(BO_3)_4$) which appear to be multiferroic materials, i.e., they can be used as magnetoelectric sensors, memory elements [10–13], etc.

Comparatively recently, great attention has been paid to orthoborate crystals co-doped with Er and Yb is associated with their potential as efficient active media solid-state lasers emitting in the spectral range 1.5–1.6 μm [14,15]. Due to high phonon frequencies (more than 1000 cm^{-1}), efficient energy transfers from Yb to Er ions take place in these crystals that is one of the crucial conditions for efficient laser action in Er-Yb co-doped materials. First of all, the laser sources in this spectral range are of great interest because of the several reasons: (1) Their emission is eye-safe since it is absorbed by cornea and does not reach retina; (2) it has low losses in atmosphere and quartz fibers; (3) room temperature sensitive detectors exist in this spectral range. Diode-laser pumping with high brightness and efficiency and long lifetime implies opportunities for the development of compact laser sources with unprecedented out parameters in different modes of operation for practical applications. Mode-locked lasers emitting in the spectral range 1.5–1.6 μm with high repetition rate are especially useful as pulse generators for high bit rate optical networks.

Single crystalline thin layers of (Er,Yb):YAl$_3$(BO$_3$)$_4$, Er:YAl$_3$(BO$_3$)$_4$ and Yb:YAl$_3$(BO$_3$)$_4$ on the undoped borate substrates also are of great interest due to their device potential. Because of the difference in the refractive index of thin film and substrate, grown epilayer exhibits waveguide properties. Potential applications of active waveguides are systems of integrated optics for high-speed signal processing.

Thus, borate crystals with huntite type structure including their derivatives are attractive for different technological applications because of their favorable physical and chemical properties like stability, high transparency, high thermal coefficient, and in particular a very high non-linear optical coefficient, making it the ideal active medium for realizing self-doubling diode pumped solid-state lasers. Wide isomorphous substitutions in R positions make it possible to extend new generation functional devices based on these solids.

In this Special Issue, different aspects of multifunctional borate materials are discussed: from ortho- and oxyorthoborates to compounds with condensed anions and from their nonlinear optical and laser properties to piezoelectric characteristics. For example, J. Dawes and coworkers investigated liquid-phase epitaxial growth of the neodymium-doped YAl$_3$(BO$_3$)$_4$ optical waveguides as potential active sources for planar integrated optics [16]. E. Cavalli and N. Leonyuk also analyzed the emission properties of the same orthoborate family [17]. Selected excitation, emission, and decay profile of rare earth-doped YAl$_3$(BO$_3$)$_4$ crystals were measured and compared with those of the concentrated compounds. The effects of the energy transfer processes and the lattice defects, as well as the ion-lattice interactions are considered taking into account the experimental results. J. Buchen at al. compared twinning in YAl$_3$(BO$_3$) and K$_2$Al$_2$B$_2$O$_7$ crystals, which may degrade crystal quality and affect nonlinear optical properties [18]. Space-resolved measurements of the optical rotation related to the twin structure were made, in order to compare the quality of these ortho- and polyborate crystals to select twin-free specimens. The piezoelectric ringing phenomenon in Pockels cells based on the beta barium borate crystals was analyzed by G. Sinkevicius and A. Baskys [19]. It was estimated that piezoelectric ringing in this metaborate crystal occurred at the 150, 205, 445, 600, and 750 kHz frequencies of high voltage pulses. F. Chan et al. also reported single crystal growth and electro-elastic properties of α-BiB$_3$O$_6$ and Bi$_2$ZnB$_2$O$_7$ crystals with the largest effective piezoelectric coefficients being in the order of 14.8 pC/N and 8.9 pC/N, respectively [20]. Finally, G. Kuzmicheva et. al. reviewed structural aspects and crystallochemical design of orthoborates belonging to huntite-type family [21]. Particular attention was paid to methods and conditions for crystal growth, affecting a crystal real composition and symmetry. A critical analysis of literature data made it possible to formulate unsolved problems in the materials science of rare-earth orthoborates, mainly scandium borates, which are distinguished by an ability to form internal and substitutional (lanthanide and Sc atoms), unlimited and limited solid solutions depending on the topological factor.

Complex investigation of phase formations in multi-component borate melts and the study of crystal growth conditions for novel high-temperature borates will provide a scientific base for development of growth technologies with device potential. On the other hand, investigations of crystal "conditions–composition–structure–properties" relationships in complex borate melts with anion polymerizations can help to create a physico-chemical base for crystal growth technology of high performance electronic and optical devices and components with a variety of industrial, medical, and entertainment applications. In the meantime, these relationships can help to estimate an affinity of synthetic borate materials with their natural prototypes and structural analogs.

The structural stability of many silicates, phosphates, and germanates also depends on the delocalization of formal charges of the A_nO_m (A = Si,Ge,P) anions as a result of their polymerization. The regular variations in their structural motifs make it possible to forecast (optimistically, more or less) new phase systems for the synthesis of advanced materials as well, because currently most of these single crystals are not available in good size or quality. A further analysis of these inorganic polymer structures will set out judicious ways towards a better understanding of the growth mechanisms of multifunctional crystals, and this Special Issue is intended to fill this gap in the field.

Acknowledgments: The Guest Editor thanks all the authors who made this Special Issue possible and the *Crystals* publishing staff for their assistance.

References

1. Franken, P.A.; Hill, A.E.; Peters, C.W.; Weinreich, G. Generation of Optical Harmonics. *Phys. Rev. Lett.* **1961**, *7*, 118–119. [CrossRef]
2. Monchamp, R.R. The distribution coefficient on neodymium and lutetium in Czochralski grown $Y_3Al_5O_{12}$. *J. Cryst. Growth* **1971**, *11*, 310–312. [CrossRef]
3. *Inorganic Crystal Structure Data Base—ICSD*; Fachinformations Zentrum (FIZ) Karlsruhe: Karlsruhe, Germany; Available online: http://icsd.fiz-karlsruhe.de/ (accessed on 3 March 2019).
4. Chen, C.; Wu, Y.; Li, R. The development of new NLO crystals in the borate series. *J. Cryst. Growth* **1990**, *99*, 790–798. [CrossRef]
5. Leonyuk, N.I.; Leonyuk, L.I. Growth and characterization of $RM_3(BO_3)_4$ crystals. *Prog. Cryst. Growth Charact. Mater.* **1995**, *31*, 179–278. [CrossRef]
6. Leonyuk, N.I. Half a century of progress in crystal growth of multifunctional borates $RAl_3(BO_3)_4$ (R = Y, Pr, Sm-Lu). *J. Cryst. Growth* **2017**, *476*, 69–77. [CrossRef]
7. Gorbel, G.; Leblanc, M.; Antic-Fidancev, E.; Lamaitre-Blaise, M.; Krupa, J.C. Luminescence analysis and subsequent revision of the crystal structure of triclinic L-$EuBO_3$. *J. Alloy. Compd.* **1999**, *287*, 71–78. [CrossRef]
8. Boyer, D.; Bertrand-Chadeyron, G.; Mahiou, R.; Lou, L.; Brioude, A.; Mugnier, J. Spectral properties of $LuBO_3$ powders and thin films processed by the sol-gel technique. *Opt. Mater.* **2001**, *16*, 21–27. [CrossRef]
9. Wei, Z.G.; Sun, L.D.; Liao, C.S.; Jiang, X.C.; Yan, C.H. Synthesis and size dependent luminescent properties of hexagonal $(Y,Gd)BO_3$:Eu nanocrystals. *J. Mater. Chem.* **2002**, *12*, 3665–3670. [CrossRef]
10. Zvezdin, A.K.; Vorob'ev, G.P.; Kadomtseva, A.V.; Popov, Y.F.; Pyatakov, A.P.; Bezmaternykh, L.N.; Kuvardin, A.V.; Popova, E.A. Magnetoelectric and magnetoelastic interactions in $NdFe_3(BO_3)_4$ multiferroics. *JETP Lett.* **2006**, *83*, 509–514. [CrossRef]
11. Begunov, A.I.; Demidov, A.A.; Gudim, I.A.; Eremin, E.V. Features of the magnetic and magnetoelectric properties of $HoAl_3(BO_3)_4$. *JETP Lett.* **2013**, *97*, 528–534. [CrossRef]
12. Kadomtseva, A.M.; Popov, Y.F.; Vorob'ev, G.P.; Kostyuchenko, N.V.; Popov, A.I.; Mukhin, A.A.; Ivanov, V.Y.; Bezmaternykh, L.N.; Gudim, I.A.; Temerov, V.L.; et al. High-temperature magnetoelectricity of terbium aluminum borate: The role of excited states of the rare-earth ion. *Phys. Rev. B* **2014**, *89*, 014418. [CrossRef]
13. Bludov, A.N.; Savina, Y.O.; Pashchenko, V.A.; Gnatchenko, S.L.; Maltsev, V.V.; Kuzmin, N.N.; Leonyuk, N.I. Magnetic properties of a $GdCr_3(BO_3)_4$ single crystal. *Low Temp. Phys.* **2018**, *44*, 423–427. [CrossRef]
14. Tolstik, N.A.; Kisel, V.E.; Kuleshov, N.V.; Maltsev, V.V.; Leonyuk, N.I. Er,Yb:$YAl_3(BO_3)_4$—Efficient 1.5 μm laser crystal. *Appl. Phys. B* **2009**, *97*, 357–362. [CrossRef]
15. Lagatsky, A.A.; Sibbett, W.; Kisel, V.E.; Troshin, A.E.; Tolstik, N.A.; Kuleshov, N.V.; Leonyuk, N.I.; Zhukov, A.E.; Rafailov, E.U. Diode-pumped passively mode-locked Er,Yb:$YAl_3(BO_3)_4$ laser at 1.5–1.6 μm. *Opt. Lett.* **2008**, *33*, 83–85. [CrossRef] [PubMed]
16. Lu, Y.; Dekker, P.; Dawes, J.M. Liquid-Phase Epitaxial Growth and Characterization of Nd:$YAl_3(BO_3)_4$ Optical Waveguides. *Crystals* **2019**, *9*, 79. [CrossRef]
17. Cavalli, E.; Leonyuk, N.I. Comparative Investigation on the Emission Properties of $RAl_3(BO_3)_4$ (R = Pr, Eu, Tb, Dy, Tm, Yb) Crystals with the Huntite Structure. *Crystals* **2019**, *9*, 44. [CrossRef]

18. Buchen, J.; Wesemann, V.; Dehmelt, S.; Gross, A.; Rytz, D. Twins in $YAl_3(BO_3)_4$ and $K_2Al_2B_2O_7$ Crystals as Revealed by Changes in Optical Activity. *Crystals* **2019**, *9*, 8. [CrossRef]
19. Sinkevicius, G.; Baskys, A. Investigation of Piezoelectric Ringing Frequency Response of Beta Barium Borate Crystals. *Crystals* **2019**, *9*, 49. [CrossRef]
20. Chen, F.; Cheng, X.; Yu, F.; Wang, C.; Zhao, X. Bismuth-Based Oxyborate Piezoelectric Crystals: Growth and Electro-Elastic Properties. *Crystals* **2019**, *9*, 29. [CrossRef]
21. Kuz'micheva, G.M.; Kaurova, I.A.; Rybakov, V.B.; Podbel'skiy, V.V. Crystallochemical Design of Huntite-Family Compounds. *Crystals* **2019**, *9*, 100. [CrossRef]

crystals

MDPI

Article

Liquid-Phase Epitaxial Growth and Characterization of Nd:YAl$_3$(BO$_3$)$_4$ Optical Waveguides

Yi Lu, Peter Dekker and Judith M. Dawes *

MQ Photonics, Department of Physics and Astronomy, Macquarie University, 2109 Sydney, Australia; coeus.lu@gmail.com (Y.L.); peter.dekker@mq.edu.au (P.D.)
* Correspondence: judith.dawes@mq.edu.au; Tel.: +61-298-508-903

Received: 16 December 2018; Accepted: 28 January 2019; Published: 1 February 2019

Abstract: We investigated the fabrication of neodymium doped thin film optical waveguide-based devices as potential active sources for planar integrated optics. Liquid-phase epitaxial growth was used to fabricate neodymium-doped yttrium aluminum borate films on compatible lattice-matched, un-doped yttrium aluminum borate substrates. We observed the refractive index contrast of the doped and un-doped crystal layers via differential interference contrast microscopy. In addition, characterization by X-ray powder diffraction, optical absorption and luminescence spectra demonstrated the crystal quality, uniformity and optical guiding of the resulting thin films.

Keywords: thin film crystal growth; epitaxial layer growth; multifunctional borate crystals; planar optical waveguides

---

1. Introduction

Integrated optics and photonics are increasingly important for optical signal processing in many applications. They rely on the development of compact, robust planar optical devices. Integrated optics employ waveguides as the building blocks for optical components, which are then connected into circuits [1]. In particular, integrated optical systems using waveguide-based components typically include active devices such as lasers and modulators that are integrated into photonic circuits. In each case, the structure must be designed to guide and confine the light within the active region of the device via the careful design of the refractive index contrast between the cladding and the active layer. Optical waveguides are described as single mode or multimode, based on the properties of the structure, as determined by the refractive index and dimensions of the guiding layer and substrate or cladding [2]. Active waveguide devices offer particular advantages over their bulk counterparts, because the optical confinement increases the intensity of the signal and pump light within the waveguide, hence ensuring that the amplification or nonlinear optical frequency conversion is more efficient than comparable bulk devices [3]. Here, we consider multimode active planar dielectric devices.

Fabrication of planar waveguide devices has been accomplished by a variety of approaches [3]. For example, the refractive index inside a dielectric material may be modified using nonlinear multiphoton processes [4,5], or ion-exchange [6]. In another approach, two crystals may be optically polished and then thermally bonded to achieve strong adhesion between the crystal layers [7]. This typically requires careful polishing to ensure that the active layer is sufficiently flat and thin. Epitaxial growth processes [8], such as liquid phase epitaxy (LPE) [9], pulsed laser deposition [10], molecular beam epitaxy (MBE) [11], hydrothermal epitaxy [12], metal oxide chemical vapor deposition (MOCVD) [13], and halide vapor phase epitaxy (HVPE) [14], enable the production of high-quality crystalline materials, which are practical for waveguide fabrication.

There has been long-standing interest in multifunctional-doped borate laser crystals [15–24], which are used in compact robust lasers emitting fundamental or self-frequency-doubled wavelengths, with Q-switched, mode-locked or continuous wave operation. The thermal conductivity of these crystals facilitates the operation of the lasers at high power and in thin disk geometries [25–27]. In addition, un-doped borates have been adopted for nonlinear optics [28,29]. This has led to a new drive for improved growth techniques for these crystals. Various approaches to optimize the crystal growth—the choice, preparation, mixing of the flux, and the temperature profile—have been reported [15–17]. There is a balance between the relatively slow growth of the crystals, and the control of the crystal phase and uniformity, due to the formation of crystal twins [30,31]. However, following an early report of epitaxial film growth [32], there has been recent interest in developing borate crystals for waveguide devices that are compatible with integrated optics. This geometry enables the concentration of the light in the active layer to enhance both the amplification and the optical nonlinearity of the device [33–36].

The liquid phase epitaxial growth technique has several advantages compared with other waveguide fabrication techniques. The layers are grown isothermally with homogeneous composition, so the quality of the epitaxial layer is comparable with that of bulk materials. The interface between the thin film and the substrate exhibits a step profile in the refractive index, whereas other waveguide fabrication techniques typically lead to graded index profiles. Generally, the modes propagating inside a multimode step index profile structure have a uniform effective index, whereas for a graded index profile, different modes have a different effective index. The thickness of the waveguide structure can be controlled accurately by the growth duration and growth temperature. Finally, liquid phase epitaxy is adaptable for any single crystalline layer or active dopant, using an appropriate flux system and growth conditions [8].

We investigated liquid phase epitaxy as an effective growth method for $Nd:YAl_3(BO_3)_4$ (Nd:YAB) thin films on compatible lattice-matched un-doped borate substrates. The resulting films, whose lattice constants are consistent with R32 crystal symmetry, exhibit very good optical properties. Differential interference contrast microscopy, X-ray powder diffraction and optical absorption and luminescence spectra were used to characterize the optical quality and uniformity of these thin films.

2. Crystal Growth

2.1. Crystal Growth Methods

The flux system for the Nd:YAB epitaxial growth was chosen to be $K_2Mo_3O_{10}$ with excess Y_2O_3 and B_2O_3 [16]. This flux system offers advantages, because it has lower volatility than the PbF_2-$3B_2O_3$ flux system, and excess Y_2O_3 and B_2O_3 were added to the initial flux to suppress Al_5BO_9 inclusions and to compensate for the volatility of B_2O_3 during crystal growth [16]. The mix for the growth was calculated as 8 at. % Nd/(Nd + Y) with 24.4 wt. % of Nd:YAB in the solution. The solvent composition was 91.9 wt. % of $K_2Mo_3O_{10}$ + 5.4 wt. % B_2O_3 + 0.25 wt. % Y_2O_3. All the chemicals were obtained from local suppliers and heat-treated in a 300 °C furnace to remove adsorbed water before weighing. They were completely ground and mixed in the platinum crucible (5 cm diameter) and heated in an electric resistance furnace at 1150 °C for 24 h. The temperature was then dropped to about 1000 °C to find the actual saturation point by repeated seeding. The seed was settled to the mid part of the solution to ensure a homogeneous temperature gradient.

The substrate for the thin film growth needs to be selected carefully. It must permit reasonable lattice matching with the epitaxial layer to avoid strain due to lattice mismatch, and it must ensure a refractive index contrast, so that the active layer is the guiding layer with a higher refractive index. We selected un-doped $YAl_3(BO_3)_4$ (YAB), as it satisfies these criteria well. We also determined that a neodymium fraction of 8% permitted waveguide confinement, without excessive lattice mismatch. The calculated refractive index contrast for 8 at % Nd dopant is 0.0632 for n_o and 0.0608 for n_e [16]. Figure 1a shows the prismatic faces {$11\bar{2}0$} and {$2\bar{1}\bar{1}0$} and the rhombic face {$01\bar{1}1$} for Nd:YAB, along with the crystal axes in Figure 1b. Previous liquid phase epitaxial growth of NdAB yielded thin films of good quality using $Gd_{0.59}La_{0.41}Al_3(BO_3)_4$ substrates with growth rates of around 1 µm/min [32].

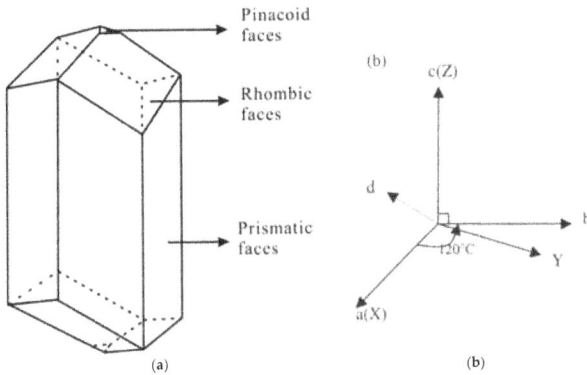

Figure 1. (a) The growth habit of Nd:YAB crystals and (b) the hexagonal crystal axes for the crystals.

The substrates were cut on the rhombic face $\{01\bar{1}1\}$ from bulk YAB crystals grown in our own laboratory and that of Professor N. Leonyuk. The surfaces of the substrates were left unpolished. The substrates were dipped vertically into the melt with platinum wire wrapped around the top of the substrates. The $\{01\bar{1}1\}$ cut pure YAB substrate (typical dimension 1 mm × 2 mm × 5 mm) was introduced and placed in the centre of the flux in the crucible.

Liquid phase epitaxial growth of thin films is typically similar to that of bulk crystals. The temperature is selected to be below the saturation point to allow the thermodynamic growth of layers with the same orientation as the substrate while immersed in a super-saturated solution. At conditions that are close to equilibrium, the deposition of the crystal on the substrate is slow and uniform. In our case, the temperature was initially set to 1 °C above the saturation point to smooth and dissolve the surface (which becomes the substrate–film interface) and the temperature was then ramped down to 4 °C below the saturation point (around 1000 °C) to start thin film growth. The growth rate was about 5 μm per hour at this temperature, as seen in Figure 2. This growth condition was chosen to ensure the thermodynamic growth regime and avoid any risk of spontaneous nucleation, which appears to occur for ΔT of 8 °C or more. After growth, the substrate was carefully removed from the flux and slowly cooled down to room temperature over 24 h. Bulk Nd:YAB crystals were also grown in our lab using the liquid phase epitaxy technique in the same solution. The solution mix was the same as for the thin film growth, with a small YAB seed.

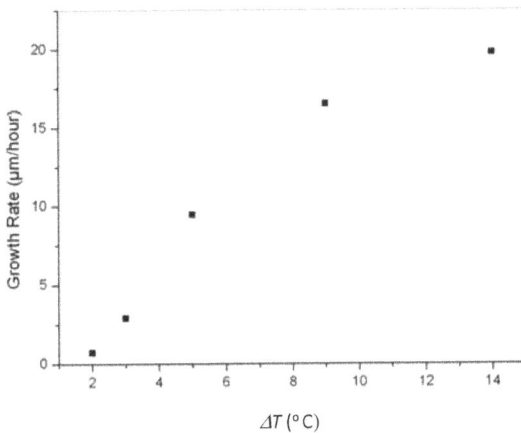

Figure 2. Nd:YAB crystal growth rate on the rhombic face seed crystal versus temperature below saturation ΔT (°C).

2.2. Results of Crystal Growth

Figure 3a shows an image of an as-grown Nd:YAB bulk crystal, with the growth facets visible, grown for two weeks under similar conditions as the thin films, with the temperature ramping down by 0.5 °C per day. Figure 3b shows an as-grown Nd:YAB crystalline thin film with un-doped YAB substrate that has been side-polished and imaged by a differential interference contrast microscope (Olympus BX60, Olympus Life Science, Sydney, Australia). The Nd:YAB thin film layer appears as a uniform, smooth blue stripe in the microscope image with a sharp change and an apparent phase contrast with the pure YAB substrate, which suggests a step-like refractive index profile. The film thickness was measured to be 71 ± 0.5 μm. The film shown in Figure 3b was obtained after epitaxial growth for 12 h at the conditions specified above. The as-grown thin film sample was transparent and homogeneous with a smooth surface. No noticeable crystallites of the monoclinic form of Nd:YAB were obtained, as discussed in Ref [35]. We attribute this to the moderate Nd dopant fraction and the lower temperature growth process that we used.

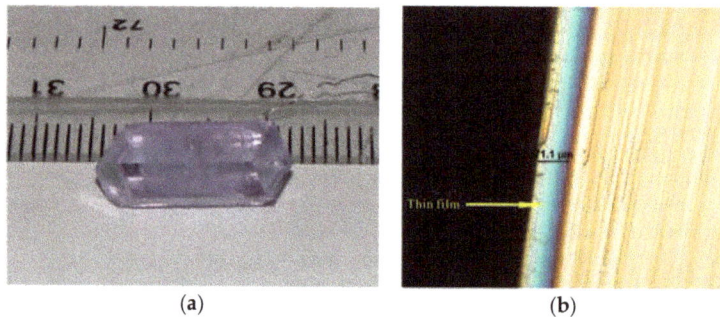

Figure 3. (**a**) As-grown bulk Nd:YAB crystal and (**b**) differential interference contrast image of the Nd:YAB thin film and YAB substrate in cross-section.

3. Crystal Characterization Methods and Results

Crystal Characterization Results

X-ray powder diffraction was used to characterise the crystallographic structure, and the results were compared with the diffraction patterns in an existing database. The X-ray powder diffraction (XRD) pattern of ground Nd:YAB bulk crystals grown by top-seeded solution growth was measured using a D/max-rA type X-ray diffractometer (Rigaku, Rigaku Americas Corp, The Woodlands, TX USA) with CuKα radiation (λ = 1.54056 Å) at room temperature, and is shown in Figure 4. The X-ray powder diffraction pattern of the Nd:YAB crystals was found to be consistent with the reference pattern of $YAl_3(BO_3)_4$ and (JCPDS card No. 15-117) [37], indicating that the neodymium dopant does not significantly perturb the lattice and the crystal belongs to the R32 space group. The lattice parameters were calculated by the least-squares method.

According to the X-ray diffraction data, the calculated lattice constants of the Nd:YAB crystals were $a = b$ = 9.298 Å and c = 7.2406 Å. In comparison with the data presented in Ref [16], the measured lattice constants of the bulk Nd:YAB crystal were close to those of $Nd_{0.09}Y_{0.91}Al_3(BO_3)_4$, which were 9.295 Å and 7.236 Å. Thus our crystal properties are consistent with 9 at. % N dopant concentration within experimental error. Lattice constants of YAB crystals with different dopants are listed in Table 1. Assuming that the thin film has a similar Nd concentration to that of the bulk crystal, the lattice mismatch of the as-grown Nd:YAB thin film is about 0.11% and 0.2% for the lattice constants a and c, respectively. From optical microscopy and visual inspection, the Nd dopant distribution in the film appeared uniform. This is expected, as the solute concentration does not vary significantly during thin film growth.

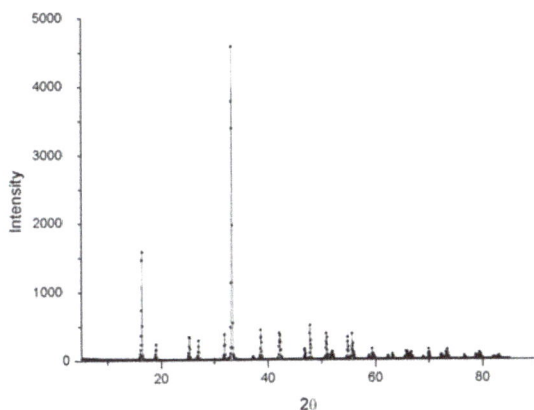

Figure 4. X-ray powder diffraction for as-grown Nd:YAB crystal.

Table 1. Lattice parameters of $YAl_3(BO_3)_4$.

Crystals	*a* (Å)	*c* (Å)	Reference
YAB (JCPDS No. 15-117)	9.2872	7.2433	[37]
$Nd_{0.09}Y_{0.91}Al_3(BO_3)_4$	9.295	7.236	[16]
$NdAl_3(BO_3)_4$	9.365	7.262	[16]
Nd:YAB crystal	9.298	7.2406	This work

The substrate crystal surface quality is shown in Figure 5a before thin film growth, and it shows the rough unpolished surface, which was smoothed in the initial period at a higher melt temperature. Figure 5b shows the as-grown crystalline film image in cross-section. The epitaxial growth results in a smooth surface with no additional crystallite formation. For the purposes of crystal characterization, the absorption spectrum of a 2 mm × 4 mm × 1.1 mm slice cut and polished from a bulk Nd:YAB crystal was measured using a Cary 5E spectrophotometer. This was compared with the absorption along the guiding direction of the epitaxial layer, as measured in the set up shown in Figure 6. Figure 7a shows the (unpolarised) absorption spectrum of bulk Nd:YAB (thickness 1.1 mm) in the wavelength range 300–1000 nm. There are six main absorption peaks in the spectrum at 360, 528, 588, 750, 809, and 882 nm, which may be assigned according to Reference [16]. The uncorrected absorption spectrum for the thin film is illustrated in Figure 7b. The absorption peak positions and features are similar to those for the bulk sample. The drifting base line is attributed to the wavelength response of the detector in the Ocean Optics HR2000 spectrometer.

(a)

(b)

Figure 5. (a) Microscope images of substrate surface (5×) and (b) after growth thin film surface (20×) for Nd:YAB on a YAB substrate. The substrate is held vertically in the flux.

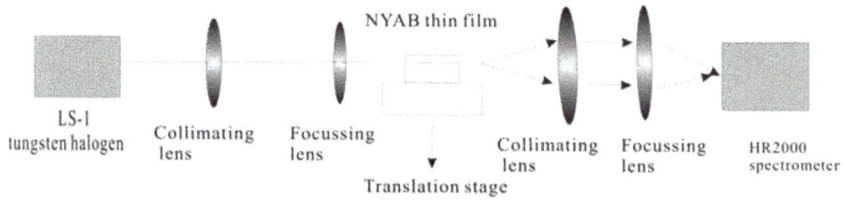

Figure 6. Experimental setup for measuring absorption spectra along guiding direction.

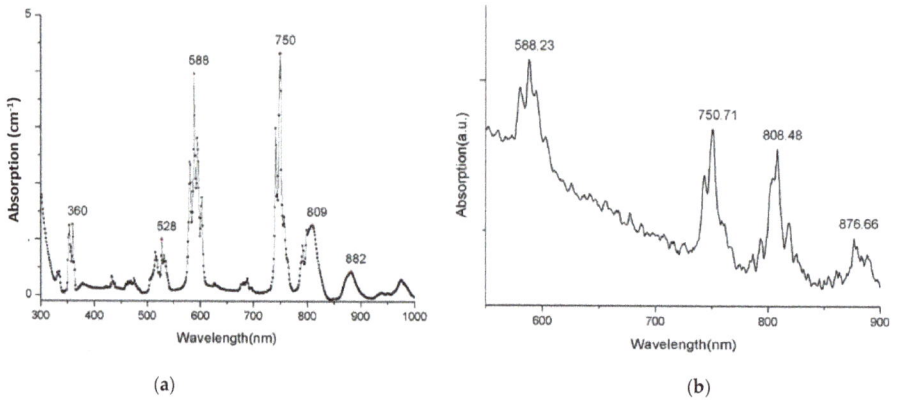

(a)

(b)

Figure 7. (a) Absorption spectrum of a polished slice of bulk Nd:YAB and (b) uncorrected absorption spectrum of Nd:YAB crystalline thin film, along the guiding direction, showing intensity peaks at 588, 750 and 808 nm, attributed to transitions from $^4I_{9/2}$ to $^4G_{5/2}$, $^4F_{7/2}$ and $^2H_{9/2}$, respectively.

The near-infrared fluorescence spectra of the Nd:YAB thin film (thickness 71 µm), and a bulk Nd:YAB crystal sample at room temperature, are overlaid in Figure 8. This spectrum was obtained by coupling light from an 808 nm diode laser into the thin film and focussing the output by a lens into a fibre-coupled spectrometer. The 887 nm, 1062 nm and 1339 nm peaks correspond to the fluorescence of the $^4F_{3/2}$ level into the $^4I_{9/2}$, $^4I_{11/2}$, and $^4I_{13/2}$ multiplets, respectively. The room temperature fluorescence peak at 1062 nm is very strong, and the bandwidth (FWHM) of the 1062 nm peak is about 10 nm. The fluorescence spectrum of the thin film Nd:YAB crystal is well-correlated with that of the bulk Nd:YAB sample.

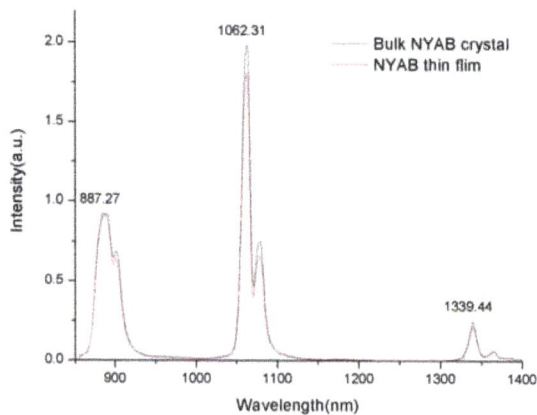

Figure 8. Luminescence spectrum of Nd:YAB thin film overlaid with that of the bulk Nd:YAB sample.

A near-field image of the Nd:YAB thin film luminescence emitted from the end of the waveguide, as longitudinally pumped by the 808 nm diode laser, is shown in Figure 9. This was captured by a CCD camera (Pulnix, JAI Pulnix, Sydney, Australia) through a 1064 nm band pass filter. The camera was coupled to a laser beam analyser (Spiricon LBA100A, Ophir-Spiricon Photonics, West North Logan, UT, USA). The luminescence image size is about 71 μm in the guided direction and 400 μm in the unguided direction. Two subsidiary images, probably due to back reflections, are also observable to the upper left. This image shows strong evidence of effective guiding within the epitaxial film layer.

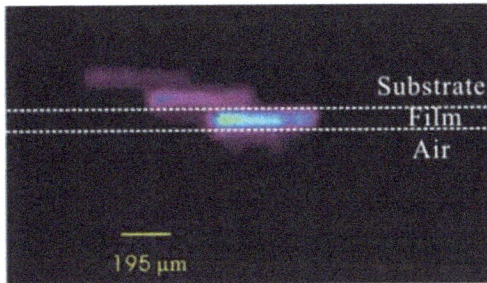

Figure 9. Luminescence image of Nd:YAB thin film (image artefacts to the upper left).

4. Discussion and Conclusions

Nd:YAB planar waveguides with 9% Nd dopant were successfully grown by the top-seeded solution method from the $K_2Mo_3O_{10}$ and B_2O_3 flux system. The growth rates were measured and the growth conditions and procedure were selected for high-quality film growth in the thermodynamic regime. Nd:YAB thin film layers were obtained at 4 °C below the saturation temperature, with a growth rate about 5 μm/h on an un-doped $\{01\bar{1}1\}$ YAB substrate. The growth of the thin films occurred over about 12 h, with additional time for cooling. The as-grown thin films have good surface and optical quality and homogeneity, and exhibit effective waveguiding of the strong luminescence at 1062 nm. Future measurements of optical gain and transmission losses in the devices are planned.

Author Contributions: Conceptualization J.M.D. and Y.L.; crystal growth methods and experiments Y.L.; crystal characterization Y.L. and P.D.; writing J.M.D. and Y.L. with editing by P.D.; and project administration J.M.D.

Funding: The authors acknowledge funding support to establish the crystal growth and characterization facility by Macquarie University and the Australian Research Council.

Acknowledgments: The authors acknowledge valuable advice and discussions on crystal growth and flux preparation from Nikolay Leonyuk and Robert Feigelson. Leonyuk provided bulk YAB crystals, which were used as substrates for the thin film growth.

Conflicts of Interest: The authors declare no conflict of interest.

References

1. Tien, P.K. Light waves in thin films and integrated optics. *Appl. Opt.* **1971**, *10*, 2395–2413. [CrossRef] [PubMed]
2. Snyder, A.W.; Love, J. *Optical Waveguide Theory*; Chapman and Hall: London, UK, 1983; pp. 6–26, ISBN 0-412-24250-8.
3. Mackenzie, J.I. Dielectric solid state planar waveguide lasers: A review. *IEEE J. Sel. Top. Quantum Electron.* **2007**, *13*, 626–637. [CrossRef]
4. Chen, F.; Vazquez de Aldana, J.R. Optical waveguides in crystalline dielectric materials produced by femtosecond-laser micromachining. *Laser Photonics Rev.* **2014**, *8*, 251–275. [CrossRef]
5. Ams, M.; Marshall, G.D.; Dekker, P.; Piper, J.A.; Withford, M.J. Ultrafast laser-written active devices. *Laser Photonics Rev.* **2009**, *3*, 535–544. [CrossRef]

6. Honkanen, S.; West, B.R.; Tliniemi, S.; Madasamy, P.; Morrell, M.; Auxier, J.; Geraghty, D. Recent advances in ion exchanged glass waveguides and devices. *Phys. Chem. Glass. Eur. J. Glass Sci. Technol. B* **2006**, *47*, 110–120.

7. Brown, C.T.A.; Bonner, C.L.; Warburton, T.J.; Shepherd, D.P.; Tropper, A.C.; Hanna, D.C.; Meissner, H.E. Thermally bonded planar waveguide lasers. *Appl. Phys. Lett.* **1997**, *71*, 1139–1141. [CrossRef]

8. Capper, P.; Irvine, S.; Joyce, T. Epitaxial Crystal Growth: Methods and Materials. In *Springer Handbook of Electronic and Photonic Materials*; Springer Handbooks; Kasap, S., Capper, P., Eds.; Springer: Cham, Switzerland, 2017.

9. Ferrand, B.; Chambaz, B.; Couchaud, M. Liquid phase epitaxy: A versatile technique for the development of miniature optical components in single crystal dielectric media. *Opt. Mater.* **1999**, *11*, 101–114. [CrossRef]

10. Beecher, S.J.; Grant-Jacob, J.A.; Hua, P.; Prentice, J.J.; Eason, R.W.; Shepherd, D.P.; Mackenzie, J.I. Ytterbium-doped-garnet crystal waveguide lasers grown by pulsed laser deposition. *Opt. Mater. Express* **2017**, *7*, 1628–1633. [CrossRef]

11. Cho, A.Y. Advances in Molecular Beam Epitaxy. *J. Cryst. Growth* **1991**, *111*, 1–13. [CrossRef]

12. Terry, R.J.; McMillen, C.D.; Chen, X.; Wen, Y.; Zhu, L.; Chumanov, G.; Kolis, J.W. Hydrothermal single crystal growth and second harmonic generation of Li_2SiO_3, Li_2GeO_3, and $Li_2Si_2O_5$. *J. Cryst. Growth* **2018**, *493*, 58–64. [CrossRef]

13. Sun, H.; Li, K.H.; Torres Castanedo, C.G.; Okur, S.; Tompa, G.S.; Salagaj, T.; Lopatin, S.; Genovese, A.; Li, X. HCl Flow-Induced Phase Change of α-, β-, and ε-Ga_2O_3 Films Grown by MOCVD. *Cryst. Growth Des.* **2018**, *18*, 2370–2376. [CrossRef]

14. Tassev, V.; Vangala, S.; Peterson, R.; Kimani, M.; Snure, M.; Markov, I. Homo and heteroepitaxial growth and study of orientation-patterned GaP for nonlinear frequency conversion devices. *Proc. SPIE* **2016**, *9731*, 97310G, "Nonlinear Frequency Generation and Conversion: Materials, Devices, and Applications XV, (4 March 2016); Vodopyanov, K.L.; Schepler, K.L.; Eds". [CrossRef]

15. Ballman, A.A. A new series of synthetic borates isostructural with the carbonate mineral huntite. *Am. Mineral.* **1962**, *47*, 1380–1383.

16. Leonyuk, N.I.; Leonyuk, L.I. Growth and characterization of $RM_3(BO_3)_4$ crystals. *Prog. Cryst. Growth Charact.* **1995**, *31*, 179–278. [CrossRef]

17. Wang, P.; Dawes, J.M.; Dekker, P.; Knowles, D.S.; Piper, J.A.; Lu, B.S. Growth and evaluation of ytterbium doped yttrium aluminum borate as a potential self-doubling laser crystal. *J. Opt. Soc. Am. B* **1999**, *16*, 63–69. [CrossRef]

18. Jaque, D.; Enguita, O.; Garcia Sole, J.; Jiang, A.D.; Luo, Z.D. Infrared continuous wave laser gain in neodymium aluminum borate: A promising candidate for microchip diode-pumped solid-state lasers. *Appl. Phys. Lett.* **2000**, *76*, 2176–2178. [CrossRef]

19. Wang, P.; Dekker, P.; Dawes, J.M.; Piper, J.A.; Liu, Y.G.; Wang, J.Y. Efficient continuous wave self-frequency doubling green diode pumped Yb:YAB lasers. *Opt. Lett.* **2000**, *25*, 731–733. [CrossRef] [PubMed]

20. Dekker, P.; Dawes, J.M.; Piper, J.A.; Liu, Y.G.; Wang, J.Y. 1.1W CW self-frequency doubled diode-pumped Yb:$YAl_3(BO_3)_4$ laser. *Opt. Commun.* **2001**, *195*, 431–436. [CrossRef]

21. Lederer, M.J.; Hildebrandt, M.; Kolev, V.Z.; LutherDavies, B.; Taylor, B.; Dawes, J.M.; Dekker, P.; Piper, J.A.; Tan, H.H.; Jagadish, C. Passive mode locking of a self-frequency doubling Yb:$YAl_3(BO_3)_4$ laser. *Opt. Lett.* **2002**, *27*, 436–438. [CrossRef]

22. Dekker, P.; Dawes, J.M.; Piper, J.A. 2.27 W Q-switched self-doubling Yb:YAB laser with controllable pulse length. *J. Opt. Soc. Am. B* **2005**, *22*, 378–384. [CrossRef]

23. Chen, Y.J.; Lin, Y.F.; Gong, X.H.; Tan, Q.G.; Luo, Z.D.; Huang, Y.D. 2.0 W diode-pumped Er,Yb:$YAl_3(BO_3)_4$ efficient 1.5 m laser crystal. *Appl. Phys. Lett.* **2006**, *89*, 241111. [CrossRef]

24. Lagatsky, A.A.; Sibbett, W.; Kisel, V.E.; Troshin, A.E.; Tolstik, N.A.; Kuleshov, N.V.; Leonyuk, N.L.; Zhukov, A.E.; Rafailov, E.U. Diode-pumped passively mode locked Er,Yb:$YAl_3(BO_3)_4$ laser at 1.5–1.6 m. *Opt. Lett.* **2008**, *33*, 83–85. [CrossRef] [PubMed]

25. Blows, J.L.; Dekker, P.; Wang, P.; Dawes, J.M.; Omatsu, T. Thermal lensing measurements and thermal conductivity of Yb:YAB. *Appl. Phys. B* **2003**, *76*, 289–292. [CrossRef]

26. Liu, J.; Mateos, X.; Zhang, H.J.; Li, J.; Wang, J.Y.; Petrov, V. High power laser performance of Yb:$YAl_3(BO_3)_4$ crystals cut along the crystallographic axes. *IEEE J. Quantum Electron.* **2007**, *43*, 385–390. [CrossRef]

27. Weichelt, B.; Rumpet, M.; Voss, A.; Wesemann, V.; Rytz, D.; Abdou Ahmed, M.; Graf, T. Yb:YAl$_3$(BO$_3$)$_4$ as gain material in thin disk oscillators: Demonstration of 109 W of IR output power. *Opt Expr.* **2013**, *21*, 25709. [CrossRef] [PubMed]

28. Yu, X.; Yue, Y.; Yao, J.; Hu, Z.G. YAl$_3$(BO$_3$)$_4$ crystal growth and characterization. *J. Cryst. Growth* **2010**, *312*, 3029–3033. [CrossRef]

29. Yu, J.; Liu, L.; Zhai, N.; Zhang, X.; Wang, G.; Wang, X.; Chen, C. Crystal growth and optical properties of YAl$_3$(BO$_3$)$_4$ for UV applications. *J. Cryst. Growth* **2012**, *341*, 61–65. [CrossRef]

30. Dekker, P.; Dawes, J.M. Characterisation of nonlinear conversion and crystal quality in Nd and Yb doped YAB. *Opt. Expr.* **2004**, *12*, 5922–5930. [CrossRef]

31. Dekker, P.; Dawes, J.M. Twinning and natural quasi-phase matching in Yb:YAB. *Appl. Phys. B* **2006**, *83*, 267. [CrossRef]

32. Lutz, F.; Leiss, M.; Muller, J. Epitaxy of NdAl$_3$(BO$_3$)$_4$ for thin film miniature lasers. *J. Cryst. Growth* **1979**, *47*, 130–132. [CrossRef]

33. Volkova, E.; Leonyuk, N. Growth of Yb:YAl$_3$(BO$_3$)$_4$ thin films by liquid-phase epitaxy. *J. Cryst. Growth* **2005**, *275*, e2467–e2470. [CrossRef]

34. Volkova, E.; Markin, V.; Leonyuk, N.I. High temperature growth and characterization of (Er,Yb):YAl$_3$(BO$_3$)$_4$ single crystal layers. *J. Cryst. Growth* **2017**, *468*, 258–261. [CrossRef]

35. Volkova, E.; Maltsev, V.; Kolganova, O.; Leonyuk, N.I. High temperature growth and characterization of Er,Yb:YAl$_3$(BO$_3$)$_4$ and NdAl$_3$(BO$_3$)$_4$ epitaxial layers. *J. Cryst. Growth* **2014**, *401*, 547–549. [CrossRef]

36. Tolstik, N.; Heinrich, S.; Kahn, A.; Volkova, E.; Maltsev, V.; Kuleshov, V.; Huber, G.; Leonyuk, N.I. High temperature growth and spectroscopic characterization of Er,Yb:YAl$_3$(BO$_3$)$_4$ epitaxial thin layers. *Opt. Mater.* **2010**, *32*, 1377–1379. [CrossRef]

37. Mills, A.D. Crystallographic data for new rare earth borate compounds RX$_3$(BO$_3$)$_4$. *Inorg. Chem.* **1962**, *1*, 960–961. [CrossRef]

crystals

MDPI

Article

Investigation of Piezoelectric Ringing Frequency Response of Beta Barium Borate Crystals

Giedrius Sinkevicius [1,2,*] **and Algirdas Baskys** [2]

[1] Department of Material Science and Electrical Engineering, Center for Physical Sciences and Technology, Saulėtekio av. 3, LT- 10257 Vilnius, Lithuania

[2] Department of Computer Science and Communications Technologies, Vilnius Gediminas Technical University, Naugarduko st. 41, LT- 03227 Vilnius, Lithuania; algirdas.baskys@vgtu.lt

[*] Correspondence: giedrius.sinkevicius@ftmc.lt; Tel.: +370-620-81193

Received: 19 December 2018; Accepted: 15 January 2019; Published: 17 January 2019

Abstract: The piezoelectric ringing phenomenon in Pockels cells based on the beta barium borate crystals was analyzed in this work. The investigation results show that piezoelectric ringing is caused by multiple high voltage pulses with a frequency in the range from 10 kHz up to 1 MHz. Experimental investigation of frequency response and Discrete Fourier transformation was used for analysis. The method of piezoelectric ringing investigation based on the analysis of difference of real and simulated optical signals spectrums was proposed. The investigations were performed for crystals with $3 \times 3 \times 25$ mm, $4 \times 4 \times 25$ mm and $4 \times 4 \times 20$ mm dimensions. It was estimated that piezoelectric ringing in the beta barium borate crystal with dimensions of 3×3 mm $\times 25$ mm occurred at the 150, 205, 445, 600 and 750 kHz frequencies of high voltage pulses.

Keywords: Pockels cell; piezo-electric ringing; beta barium borate

1. Introduction

A wide variety of optical modulators are used in laser applications [1–7]. However, only modulators based on electro-optic and acousto-optic principles are used for high power laser system applications. The electro-optic modulation principle is based on the Pockels effect, which presents the change of refractive index in non-centrosymmetric crystals under the influence of an external electric field. The change in refractive index induces a polarization change of a beam that travels though the crystal of the Pockels cell. This feature allows us to use the Pockels cell as a voltage controlled half-wave plate [8].

Acousto-optic modulation principle is based on Debye-Sears effect combined with Bragg configuration. The mechanical oscillation of piezoelectric wafer on the side of the crystal creates pressure, which increase the refraction, diffraction and interference inside the crystal [8,9]. This effect allows us to employ the acousto-optic modulator as a beam deflector.

The property of any optical modulator is the ability to let through or shut out the laser beam that travels through the modulator. The measure of this property for the electro-optic modulators is contrast ratio; for the acousto-optic modulators - diffraction efficiency. Electro-optical modulators contrast ratio varies from 1000:1 to 2000:1 [10–12] and acousto-optic modulators diffraction efficiency from 30 % to 80 % [4,13–15].

An essential parameter of optical modulators is the rise and the fall time of the modulator. The rise and fall time of electro-optical modulators can reach 10 ns [4,12] and it varies from 200 ns down to 56 ns for acousto-optic modulators [14,15]. It should be noted that the rise and fall time of the acousto-optic modulator depends on the sound velocity in the crystal and the waist diameter of the beam [14]. The rise and fall time of an electro-optical modulators based on the Pockels cell is defined by the rise and fall time of the high voltage pulse that is applied to the modulator. The required amplitude of the pulse for operation of any type of Pockels cell depends on the material, dimensions of the crystal and wavelength of the laser beam that travels through the Pockels cell crystal [16,17].

There are two types of Pockels cells: transverse and longitudinal [8]. For longitudinal Pockels cells, half-wave voltage (amplitude of pulse, at which the laser beam polarization has been rotated by 90°) may vary from 4kV to 10 kV [18]. For transverse Pockels cells, half-wave voltage varies between 1 kV and 8 kV [19,20]. Pockels cell rise/fall time depends on two factors: capacitance of the crystal and applied voltage rise/fall time. In order to achieve a shorter duration of rise/fall time of the Pockels cell, it is necessary to decrease the values of these parameters. The capacitance is determined by the dimensions and material of the crystal; therefore, the main way for the reduction of the rise/fall time of Pockels cell is improvement of the electronics of the high voltage driver. There are only a few usable concepts for high voltage driver design: metal-oxide-semiconductor field-effect transistors (MOSFET) or bipolar avalanche transistors connected in series [21]. The high voltage drivers based on the MOSFETs connected in series provide the laser beam modulation frequency up to 1 MHz with the 30 ns rise/fall time of pulse and adjustable pulse width [22–27]. The bipolar avalanche high voltage drivers allow us to achieve rise time from 7 ns down to 240 ps [28–35]. However, the fall time of the pulse generated by the bipolar avalanche high voltage drivers may be 10 times longer than the rise time of the pulse. Another drawback is that practically, it is not possible to adjust the pulse width. The highest width of high voltage pulse is around 7 ns [35] and the highest achieved frequency is 200 kHz [28].

Pockels cells are characterized by higher contrast ratio and shorter duration of rise/fall time in comparison with the acousto-optic modulators. However, the high voltage pulses with the short rise/fall duration can cause piezoelectrical ringing that induces acoustic waves in Pockels cell crystal. If acoustic waves are not suppressed, they are reflected back inside of the Pockels cell crystal. This phenomenon introduces the elasto-optic effect, which can cause the reduction of Pockels cells contrast ratio [11,36–39]. This effect has to be investigated in order to find the ways how to reduce the impact of the piezoelectric ringing on the contrast ratio. The beta barium borate (BBO) crystals are usually used for the implementation of Pockels cells. There are many works dedicated to the analysis of effects in BBO crystals, e.g., [11,40–43]. However, the piezoelectric ringing phenomenon in Pockels cells with BBO crystals was mentioned in merely one research [11] and there are none dedicated to the analysis of high voltage pulse frequency impact on piezoelectric ringing phenomenon.

The investigation results of piezoelectric ringing of the Pockels cells with BBO crystals that operate in high voltage pulse frequency range from 10 kHz up to 1 MHz are presented. The method of the piezoelectric ringing investigation based on the spectrum analysis of difference of real and simulated optical signals is proposed. The investigations are performed for the crystals with $3 \times 3 \times 25$ mm, $4 \times 4 \times 25$ mm and $4 \times 4 \times 20$ mm dimensions.

2. Investigation Procedure

Pockels cells based on the BBO crystals were investigated experimentally. A block diagram of simplified single pass contrast ratio measurement setup is given in Figure 1. A 1064 nm wavelength diode-pumper solid state (DPSS) laser with 1.5 mm beam diameter and 2 W power was used in this setup. Half-wave plate with antireflective (AR) coating at 1064 nm was used to alter beam polarization of the DPSS laser. Glan-Taylor polarizing prism was selected for its high extinction ratio, which is higher than 100,000:1 [44]. This prism is used in order to let through only the p-polarized beam. The optical attenuator in combination with half-wave plate was employed in order to adjust the power and to avoid the saturation of photodetector. Laser beam from the first Glan-Taylor polarizing prism travels through the Pockels cell and after that to the second Glan-Taylor prism. The light intensity was detected via a high-speed Si photodetector (DET10A/M, Thorlabs, Newton, NJ, United States, 2013). It must be noted that the phases of the first and the second Glan-Taylor polarizing prisms were mutually turned by 90°, i.e., the second Glan-Taylor polarizing prism was used as an analyzer.

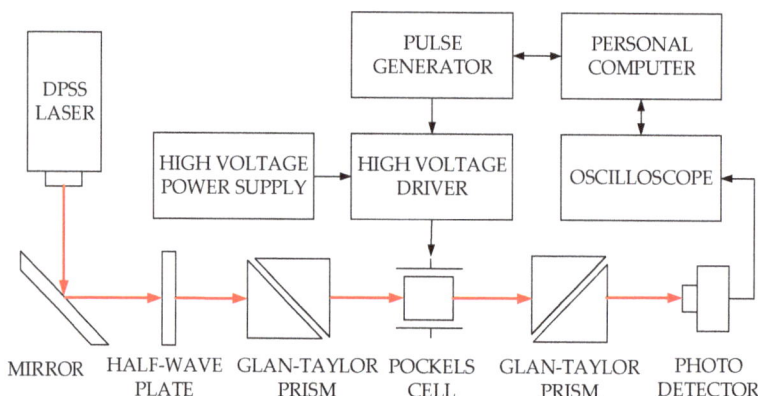

Figure 1. Single pass contrast ratio measurement setup, where DPSS - diode-pumper solid state.

Bipolar high voltage power supply (PS2-60-1.4, Eksma Optics, Vilnius, Lithuania, 2015) was used. It provided positive and negative output voltage 1 kV to 1.4 kV. A potential difference of 2.8 kV was possible with this device. Voltage was supplied to the bipolar high voltage driver (DPD-1000-2.9-Al, Eksma Optics, Vilnius, Lithuania, 2015), which produced high voltage pulses and operated on the MOSFET connected in series principle. Pulse width adjustment from 100 ns up to 5 μs was achieved with this driver model. High voltage pulse rise/fall edge duration from 8 ns down to 4 ns was achievable at pulse frequencies up to 1 MHz. Frequency and duration of the high voltage pulses was assigned via pulse generator (9530, Quantum composers, Bozeman, MT, United States, 2014).

High voltage pulses from bipolar high voltage driver were supplied to the Pockels cell. When the voltage was applied to the Pockels cell, the output signal of the photodetector became high. When no voltage was applied to the Pockels cell, photodetector output signal was low. Oscilloscope sampled the voltage on the output of the photodetector and captured the oscillograms. The intensity of the laser beam that was recorded by the photodetector can be calculated by [45]:

$$I = I_0 \sin^2 \frac{\Gamma}{2} , \tag{1}$$

where I_0 is the intensity of incident light and Γ is phase retardation, which is calculated using equation:

$$\Gamma = \frac{2\pi L n_0^2 r_{ij} V}{\lambda d} , \tag{2}$$

where L is the crystal length, n_0 is the refractive index, r_{ij} is the electro-optic coefficient, V is the voltage applied, λ is the wavelength of the laser beam and d is the crystal thickness.

Contrast ratio was calculated using the intensity value of optical signal, which was captured by the oscilograms, applying equation [46]:

$$\frac{1}{CR} = \frac{1}{\left(\frac{I_{max}}{I_{min}}\right)} , \tag{3}$$

where:

$$I_{max} = \frac{1}{t_0 - t_{pw}} \int_{t_0}^{t_{pw}} U_d(t)dt , \tag{4}$$

$$I_{min} = \frac{1}{t_p - t_{pw}} \int_{t_{pw}}^{t_p} U_d(t)dt , \tag{5}$$

where $U_d(t)$ is the voltage of photodetector, t_0 is moment at which the optical pulse starts, t_{pw} is the width of optical pulse and t_p is the period of the pulses. I_{max} and I_{min} are the average of maximum and minimum intensity of the optical signal in the input of the photodetector.

The contrast ratio of Pockels cells was measured in the high voltage pulse frequency range 10 kHz up to 1 MHz. Oscilloscope data were sampled and the frequency was changed with the step of 3 kHz automatically via software developed by the authors.

Measurements were performed for the the BBO Pockels cell crystals with the following dimensions:

- 3 mm × 3 mm × 25 mm
- 4 mm × 4 mm × 25 mm
- 4 mm × 4 mm × 20 mm

3. Investigation Results

Photodetector signal oscilograms for Pockels cells with BBO crystals of 3 × 3 × 25 mm, 4 × 4 × 25 mm and 4 × 4 × 20 mm dimensions were captured at high voltage pulses in ranges from 10 kHz to 1 MHz with 3 kHz steps. Photodetector signal for Pockels cell with 3 mm × 3 mm × 25 mm BBO crystal is presented in Figure 2. Additionally, the simulated square wave signal of photodetector is presented in Figure 2. The simulation was performed using LabVIEW software (National Instruments, Austin, TX, United States, 2016) the model used for simulation did not take into account the piezoelectric ringing phenomenon of the Pockels cell crystal.

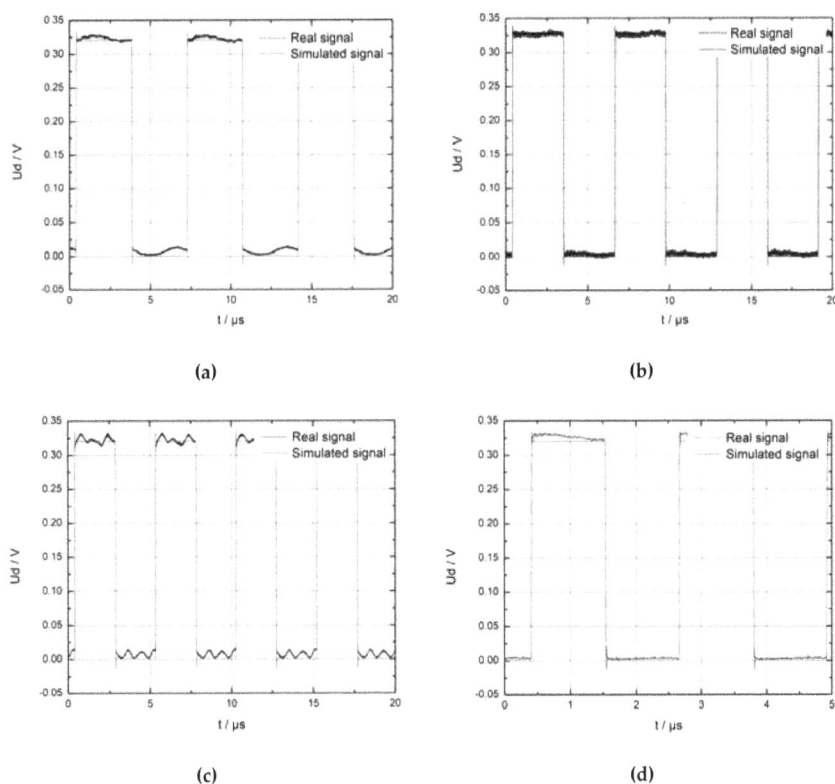

(a)

(b)

(c)

(d)

Figure 2. *Cont.*

(e)

(f)

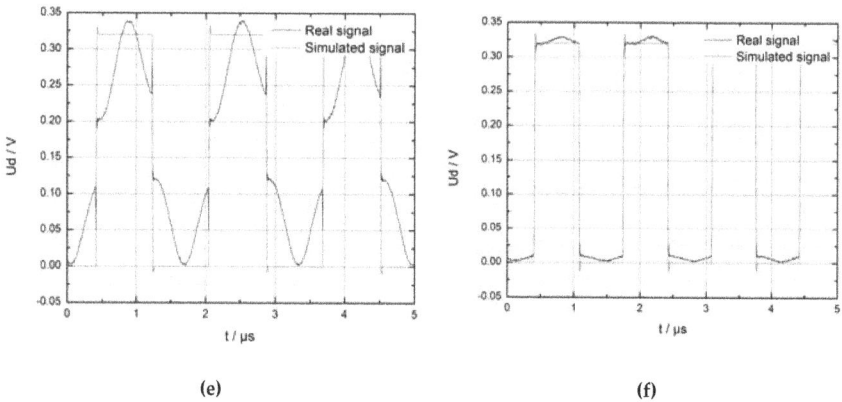

Figure 2. Real and simulated photodetector signal for Pockels cell with 3 mm × 3 mm × 25 mm beta barium borate (BBO) crystal at frequencies of high voltage pulses: (**a**) 150 kHz; (**b**) 180 kHz; (**c**) 205 kHz; (**d**) 445 kHz; (**e**) 600 kHz; (**f**) 750 kHz.

The photodetector signal was close to the simulated square wave signal, just at the 180 kHz high voltage pulse frequency (Figure 2b). This showed that the piezoelectric ringing did not occur at the 180 kHz frequency. However, in all other oscillograms that are presented in Figure 2, optical signal was distorted as compared to the simulated square wave signal. Distortion of the photodetector signal was the consequence of the piezoelectric ringing phenomenon. Oscilograms presented in Figure 2a,c–f show that the period of oscillations caused by the piezoelectric ringing phenomenon depended on the high voltage pulse frequency. The highest distortion of the optical signal was recorded at the 600 kHz high voltage pulse frequency (Figure 2e). This happens because the frequency of the piezoelectric ringing coincided with the frequency of high voltage pulses in this situation and, therefore, some resonance occurred.

The signals caused by the piezoelectric ringing oscillations at 150 kHz, 205 kHz, 445 kHz, 600 kHz and 750 kHz frequencies were analyzed using discrete Fourier transform (DFT):

$$y(f) = \sum_{n=0}^{N-1} x_n e^{-j2\pi kn/N} \tag{6}$$

where n = 0, 1, 2, ... , $N-1$, x is input sequence, N is the number of elements of x and y is the transform result.

Spectrums of real and simulated photodetector signals at the 180 kHz high voltage pulse frequency obtained using DFT are presented in Figure 3a. The difference of real and simulated photodetector signal spectrums was calculated to evaluate the spectrum inequality. The result of calculation is presented in Figure 3b. This showed that spectrums of real and simulated signals were close to each other. This fact proves that no piezoelectric ringing oscillations occurred at frequencies up to 11 MHz if the high voltage pulses with 180 kHz were used for the Pockels cell control.

The differences of real and simulated photodetector signal spectrums calculated for the signals obtained at the 150 kHz, 180 kHz, 205 kHz, 445 kHz and 750 kHz high voltage pulse frequencies are presented in Figure 4. The results for 600 kHz frequency are not displayed, since the subtraction result was very high, and therefore, the imaging of the graph would be worsened.

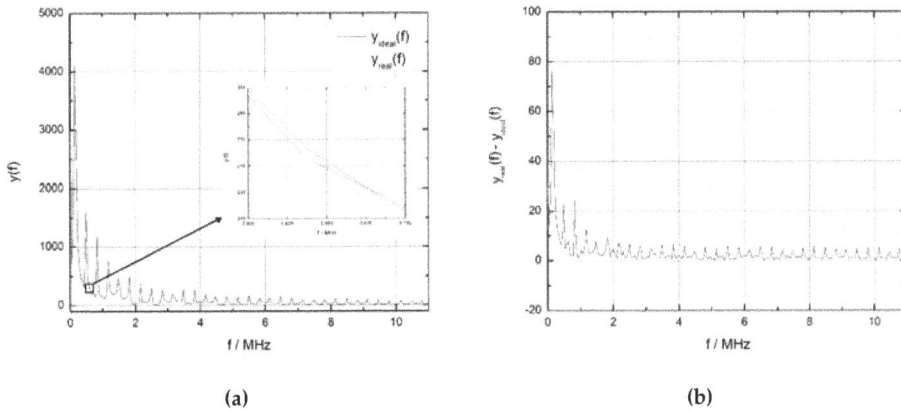

Figure 3. (**a**) Spectrums of real and simulated photodetector signals; (**b**) result of spectrum subtraction. High voltage pulse frequency at 180 kHz.

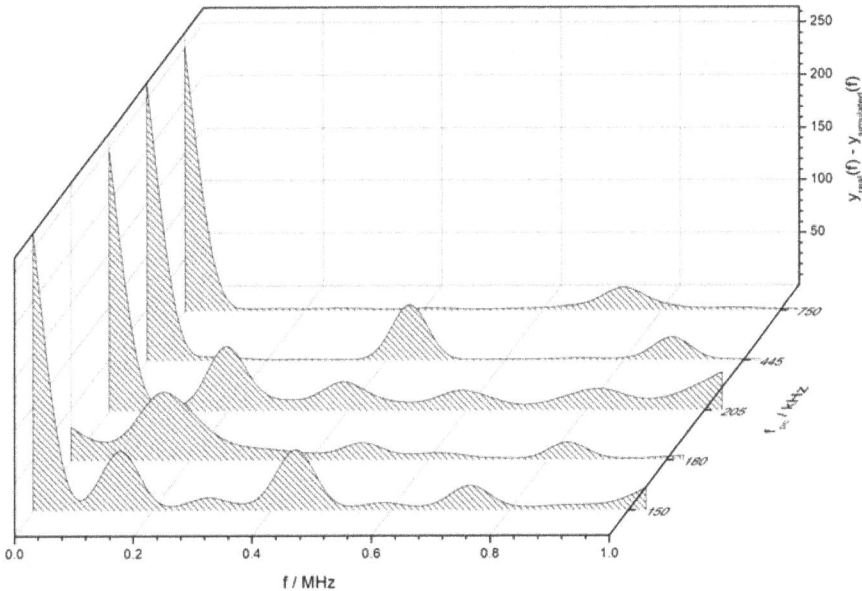

Figure 4. Real and simulated signals spectrums subtraction at 150 kHz, 180 kHz, 205 kHz, 445 kHz and 750 kHz high voltage pulse frequencies.

The highest amplitudes of spectrum components obtained for 150 kHz high voltage pulse frequency were observed at 150 kHz and 450 kHz; for 205 kHz at 205 kHz with a lot of lower amplitude components, for 445 kHz at 445 kHz and 890 kHz, and for 750 kHz at 750 kHz.

The piezoelectric ringing in the Pockels cell with BBO with dimensions of $3 \times 3 \times 25$ mm occurred if the spectrum components of the high voltage pulses were near the resonant frequencies of 445kHz, 600 kHz or 750 kHz. Therefore, the acoustic wave suppressors designed for these frequencies had to be applied.

The dependences of Pockels cell BBO crystals contrast ratio on applied high-voltage pulse frequency was obtained by processing the captured oscilograms data using the contrast ratio

Equation (3). The obtained dependences of $3 \times 3 \times 25$ and $4 \times 4 \times 25$ mm Pockels cell BBO crystals contrast on applied high-voltage pulse frequency are presented in Figure 5.

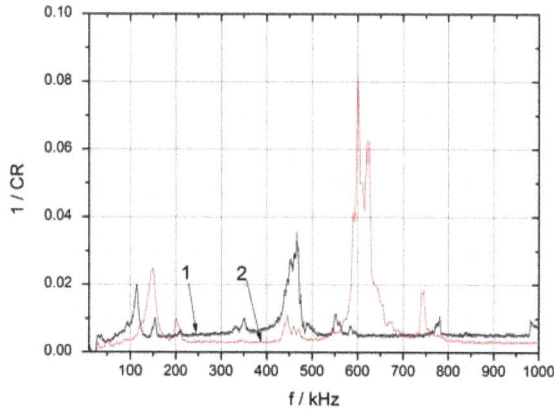

Figure 5. The dependences of Pockels cell BBO crystals contrast ratio on applied high-voltage pulse frequency for crystals with various dimensions: 1—$4 \times 4 \times 25$ mm; 2—$3 \times 3 \times 25$ mm.

The obtained results for Pockels cell with $3 \times 3 \times 25$ mm BBO crystal show that piezoelectric ringing occurred and, because of this, the contrast ratio decreased at 150 kHz, 205 kHz, 445 kHz, 600 kHz and 750 kHz high voltage pulse frequencies. Spectrum analysis shows that piezoelectric ringing at the 150 kHz and 205 kHz frequencies was caused by the third harmonic of high voltage pulses. It is seen (Figure 5) that resonant frequencies depended on the size of the BBO crystal aperture. The resonances of smaller crystal occurred at higher frequencies. When aperture of crystal changes from 3 to 4 mm, the lowest resonance frequency decreased by 36 kHz at the 114 kHz frequency, while for the highest resonance frequency it decreased by 195 kHz at the 550 kHz frequency.

It is seen that with the decreasing of BBO Pockels cell crystal length from 25 mm to 20 mm, the resonance frequency shifted slightly (Figure 6).

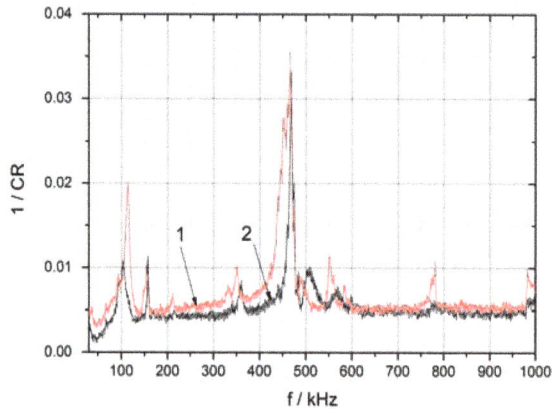

Figure 6. The dependences of Pockels cells BBO crystal contrast ratio on applied high-voltage pulse frequency for crystals with various dimensions: 1—$4 \times 4 \times 20$ mm; 2—$4 \times 4 \times 25$ mm.

4. Discussion

The novel method of piezoelectric ringing analysis of Pockels cells with BBO crystals, which is based on the subtraction of real and simulated photodetector signals spectrums, allows us to observe spectrum components caused by the piezoelectric ringing phenomenon.

The obtained results for Pockels cell with $3 \times 3 \times 25$ mm BBO crystal show that piezoelectric ringing occurred and, because of this, the contrast ratio decreased at 150 kHz, 205 kHz, 445 kHz, 600 kHz and 750 kHz high voltage pulse frequencies. Spectrum analysis shows that piezoelectric ringing at the 150 kHz and 205 kHz frequencies was caused by the third harmonic of high voltage pulses. The frequencies at which the piezoelectric ringing occurred depend on the size of the BBO crystal aperture. The piezoelectric ringing of smaller crystal occurred at higher frequencies. When aperture of crystal changed from 3 to 4 mm, the lowest frequency decreased by 36 kHz at 114 kHz frequency, while for the highest resonance frequency it decreased by 195 kHz at the 550 kHz frequency.

The highest distortion of the optical signal was recorded at the 600 kHz high voltage pulse frequency. This happened because the frequency of the piezoelectric ringing coincided with the frequency of high voltage pulses in this situation, and therefore, some resonance occurred.

The method for piezoelectric ringing suppression in Pockels cells with BBO crystals has to be developed to decrease the impact of this phenomenon on the contrast ratio of Pockels cell.

Author Contributions: G.S. and A.B. wrote the paper and analyzed the data. G.S. performed the experiments.

Funding: This research received no external funding.

Conflicts of Interest: The authors declare no conflict of interest.

References

1. Yu, T.; Gao, F.; Zhang, X.; Xiong, B.; Yuan, X. Bidirectional ring amplifier with twin pulses for high-power lasers. *Opt. Express* **2018**, *26*, 15300–15307. [CrossRef]
2. Rezvani, S.A.; Suzuki, M.; Malevich, P.; Livache, C.; de Montgolfier, J.V.; Nomura, Y.; Tsurumachi, N.; Baltuška, A.; Fuji, T. Millijoule femtosecond pulses at 1937 nm from a diode-pumped ring cavity Tm:YAP regenerative amplifier. *Opt. Express* **2018**, *26*, 29460–29470. [CrossRef]
3. Fattahi, H.; Alismail, A.; Wang, H.; Brons, J.; Pronin, O.; Buberl, T.; Vámos, L.; Arisholm, G.; Azzeer, A.M.; Krausz, F. High-power, 1-ps, all-Yb:YAG thin-disk regenerative amplifier. *Opt. Lett.* **2016**, *41*, 1126–1129. [CrossRef]
4. Giesberts, M.; Fitzau, O.; Hoffmann, H.-D.; Lange, R.; Bachert, C.; Krause, V. Directly q-switched high power resonator based on XLMA-fibers. In Proceedings of the Fiber Lasers XV: Technology and Systems; International Society for Optics and Photonics, San Francisco, CA, USA, 26 February 2018; Volume 10512, p. 1051218. [CrossRef]
5. Römer, G.R.B.E.; Bechtold, P. Electro-optic and Acousto-optic Laser Beam Scanners. *Phys. Procedia* **2014**, *56*, 29–39. [CrossRef]
6. Sun, Z.; Martinez, A.; Wang, F. Optical modulators with 2D layered materials. *Nat. Photonics* **2016**, *10*, 227–238. [CrossRef]
7. Munk, A.; Jungbluth, B.; Strotkamp, M.; Hoffmann, H.-D.; Poprawe, R.; Höffner, J. Diode-pumped Alexandrite ring laser for lidar applications. In Proceedings of the Solid State Lasers XXV: Technology and Devices; International Society for Optics and Photonics, San Francisco, CA, USA, 16 March 2016; Volume 9726, p. 97260I. [CrossRef]
8. Eichler, H.J.; Eichler, J.; Lux, O. Modulation and Deflection. In *Lasers: Basics, Advances and Applications*; Eichler, H.J., Eichler, J., Lux, O., Eds.; Springer Series in Optical Sciences; Springer International Publishing: Cham, Switzerland, 2018; pp. 299–311. ISBN 978-3-319-99895-4. [CrossRef]
9. Rashed, A.N.Z. Best candidate materials for fast speed response and high transmission performance efficiency of acousto optic modulators. *Opt. Quantum Electron.* **2014**, *46*, 731–750. [CrossRef]
10. Zhang, F.; Tang, P.; Wu, M.; Huang, B.; Liu, J.; Qi, X.; Zhao, C. Voltage-on-Type RTP Pockels Cell for Q-switch of an Er:YAG Laser at 1,617 nm. *J. Russ. Laser Res.* **2017**, *38*, 339–343. [CrossRef]

11. Bergmann, F.; Siebold, M.; Loeser, M.; Röser, F.; Albach, D.; Schramm, U. MHz Repetion Rate Yb:YAG and Yb:CaF2 Regenerative Picosecond Laser Amplifiers with a BBO Pockels Cell. *Appl. Sci.* **2015**, *5*, 761–769. [CrossRef]

12. De Groote, R.P.; Budinčević, I.; Billowes, J.; Bissell, M.L.; Cocolios, T.E.; Farooq-Smith, G.J.; Fedosseev, V.N.; Flanagan, K.T.; Franchoo, S.; Garcia Ruiz, R.F.; et al. Use of a Continuous Wave Laser and Pockels Cell for Sensitive High-Resolution Collinear Resonance Ionization Spectroscopy. *Phys. Rev. Lett.* **2015**, *115*, 132501. [CrossRef]

13. Donley, E.A.; Heavner, T.P.; Levi, F.; Tataw, M.O.; Jefferts, S.R. Double-pass acousto-optic modulator system. *Rev. Sci. Instrum.* **2005**, *76*, 063112. [CrossRef]

14. Wu, Q.; Gao, Z.; Tian, X.; Su, X.; Li, G.; Sun, Y.; Xia, S.; He, J.; Tao, X. Biaxial crystal β-BaTeMo$_2$O$_9$: Theoretical analysis and the feasibility as high-efficiency acousto-optic Q-switch. *Opt. Express* **2017**, *25*, 24893–24900. [CrossRef]

15. El-Sherif, A.F.; Harfosh, A. Comparison of high-power diode pumped actively Q-switched double-clad flower shape co-doped-Er3+:Yb3+fiber laser using acousto-optic and mechanical (optical) modulators. *J. Mod. Opt.* **2015**, *62*, 1229–1240. [CrossRef]

16. Andreev, N.F.; Babin, A.A.; Davydov, V.S.; Matveev, A.Z.; Garanin, S.G.; Dolgopolov, Y.V.; Kulikov, S.M.; Sukharev, S.A.; Tyutin, S.V. Wide-aperture plasma-electrode pockels cell. *Plasma Phys. Rep.* **2011**, *37*, 1219–1224. [CrossRef]

17. Zhang, C.; Feng, X.; Liang, S.; Zhang, C.; Li, C. Quasi-reciprocal reflective optical voltage sensor based on Pockels effect with digital closed-loop detection technique. *Opt. Commun.* **2010**, *283*, 3878–3883. [CrossRef]

18. Dorrer, C. Analysis of nonlinear optical propagation in a longitudinal deuterated potassium dihydrogen phosphate Pockels cell. *JOSA B* **2014**, *31*, 1891–1900. [CrossRef]

19. Li, C. Electrooptic Switcher Based on Dual Transverse Pockels Effect and Lithium Niobate Crystal. *IEEE Photonics Technol. Lett.* **2017**, *29*, 2159–2162. [CrossRef]

20. Ionin, A.A.; Kinyaevskiy, I.O.; Klimachev, Y.M.; Kotkov, A.A.; Kozlov, A.Y.; Kryuchkov, D.S. Selection of CO laser single nanosecond pulse by electro-optic CdTe shutter. *Infrared Phys. Technol.* **2017**, *85*, 347–351. [CrossRef]

21. Jiang, W.; Yatsui, K.; Takayama, K.; Akemoto, M.; Nakamura, E.; Shimizu, N.; Tokuchi, A.; Rukin, S.; Tarasenko, V.; Panchenko, A. Compact solid-State switched pulsed power and its applications. *Proc. IEEE* **2004**, *92*, 1180–1196. [CrossRef]

22. Rutten, T.P.; Wild, N.; Veitch, P.J. Fast rise time, long pulse width, kilohertz repetition rate Q-switch driver. *Rev. Sci. Instrum.* **2007**, *78*, 073108. [CrossRef]

23. Xu, Y.; Chen, W.; Liang, H.; Li, Y.-H.; Liang, F.-T.; Shen, Q.; Liao, S.-K.; Peng, C.-Z. Megahertz high voltage pulse generator suitable for capacitive load. *AIP Adv.* **2017**, *7*, 115210. [CrossRef]

24. Baker, R.J.; Johnson, B.P. Series operation of power MOSFETs for high-speed, high-voltage switching applications. *Rev. Sci. Instrum.* **1993**, *64*, 1655–1656. [CrossRef]

25. Jiang, W. Fast High Voltage Switching Using Stacked MOSFETs. *IEEE Trans. Dielectr. Electr. Insul.* **2007**, *14*, 947–950. [CrossRef]

26. Sundararajan, R.; Shao, J.; Soundarajan, E.; Gonzales, J.; Chaney, A. Performance of solid-state high-voltage pulsers for biological applications-a preliminary study. *IEEE Trans. Plasma Sci.* **2004**, *32*, 2017–2025. [CrossRef]

27. Keith, W.D.; Pringle, D.; Rice, P.; Birke, P.V. Distributed magnetic coupling synchronizes a stacked 25-kV MOSFET switch. *IEEE Trans. Power Electron.* **2000**, *15*, 58–61. [CrossRef]

28. Krishnaswamy, P.; Kuthi, A.; Vernier, P.T.; Gundersen, M.A. Compact Subnanosecond Pulse Generator Using Avalanche Transistors for Cell Electroperturbation Studies. *IEEE Trans. Dielectr. Electr. Insul.* **2007**, *14*, 873–877. [CrossRef]

29. Xuelin, Y.; Hongde, Z.; Yang, B.; Zhenjie, D.; Qingsong, H.; Bo, Z.; Long, H. 4kV/30kHz short pulse generator based on time-domain power combining. In Proceedings of the 2010 IEEE International Conference on Ultra-Wideband, Nanjing, China, 20–23 September 2010; Volume 2, pp. 1–4. [CrossRef]

30. Oak, S.M.; Bindra, K.S.; Narayan, B.S.; Khardekar, R.K. A fast cavity dumper for a picosecond glass laser. *Rev. Sci. Instrum.* **1991**, *62*, 308–312. [CrossRef]

31. Tamuri, A.; Bidin, N.; Mad Daud, Y. Nanoseconds Switching for High Voltage Circuit using Avalanche Transistors. *Appl. Phys. Res.* **2009**, *1*. [CrossRef]

32. Rai, V.N.; Shukla, M.; Khardekar, R.K. A transistorized Marx bank circuit providing sub-nanosecond high-voltage pulses. *Meas. Sci. Technol.* **1994**, *5*, 447. [CrossRef]

33. Dharmadhikari, A.K.; Dharmadhikari, J.A.; Adhi, K.P.; Mehendale, N.Y.; Aiyer, R.C. Low cost fast switch using a stack of bipolar transistors as a pockel cell driver. *Rev. Sci. Instrum.* **1996**, *67*, 4399–4400. [CrossRef]

34. Ding, W.; Wang, Y.; Fan, C.; Gou, Y.; Xu, Z.; Yang, L. A Subnanosecond Jitter Trigger Generator Utilizing Trigatron Switch and Avalanche Transistor Circuit. *IEEE Trans. Plasma Sci.* **2015**, *43*, 1054–1062. [CrossRef]

35. Bishop, A.I.; Barker, P.F. Subnanosecond Pockels cell switching using avalanche transistors. *Rev. Sci. Instrum.* **2006**, *77*, 044701. [CrossRef]

36. Kemp, J.C. Piezo-Optical Birefringence Modulators: New Use for a Long-Known Effect. *JOSA* **1969**, *59*, 950–954. [CrossRef]

37. Sinkevicius, G.; Baskys, A. Investigation of frequency response of pockels cells based on beta barium borate crystals. In Proceedings of the 2017 Open Conference of Electrical, Electronic and Information Sciences (eStream), Vilnius, Lithuania, 22–27 April 2017; pp. 1–4. [CrossRef]

38. Yin, X.; Jiang, M.; Sun, Z.; Hui, Y.; Lei, H.; Li, Q. Intrinsic reduction the depolarization loss in electro-optical Q-switched laser using a rectangular KD*P crystal. *Opt. Commun.* **2017**, *398*, 107–111. [CrossRef]

39. Wu, W.; Li, X.; Yan, R.; Zhou, Y.; Ma, Y.; Fan, R.; Dong, Z.; Chen, D. 100 kHz, 3.1 ns, 1.89 J cavity-dumped burst-mode Nd:YAG MOPA laser. *Opt. Express* **2017**, *25*, 26875–26884. [CrossRef]

40. Adamiv, V.T.; Ebothe, J.; Piasecki, M.; Burak, Y.V.; Teslyuk, I.M.; Plucinski, K.J.; Reshak, A.H.; Kityk, I.V. "Triggering" effect of second harmonic generation in centrosymmetric α-BaB2O4 crystals. *Opt. Mater.* **2009**, *31*, 685–687. [CrossRef]

41. Andrushchak, A.S.; Adamiv, V.T.; Krupych, O.M.; Martynyuk-Lototska, I.Y.; Burak, Y.V.; Vlokh, R.O. Anisotropy of piezo- and elastooptical effect in β-BaB2O4 crystals. *Ferroelectrics* **2000**, *238*, 299–305. [CrossRef]

42. Takahashi, M.; Osada, A.; Dergachev, A.; Moulton, P.F.; Cadatal-Raduban, M.; Shimizu, T.; Sarukura, N. Effects of Pulse Rate and Temperature on Nonlinear Absorption of Pulsed 262-nm Laser Light in β-BaB2O4. *Jpn. J. Appl. Phys.* **2010**, *49*, 080211. [CrossRef]

43. Trnovcová, V.; Kubliha, M.; Kokh, A.; Fedorov, P.P.; Zakalyukin, R.M. Electrical properties of crystalline borates. *Russ. J. Electrochem.* **2011**, *47*, 531. [CrossRef]

44. Takubo, Y.; Takeda, N.; Huang, J.H.; Muroo, K.; Yamamoto, M. Precise measurement of the extinction ratio of a polarization analyser. *Meas. Sci. Technol.* **1998**, *9*, 20. [CrossRef]

45. Roth, M.; Tseitlin, M.; Angert, N. Oxide crystals for electro-optic Q-switching of lasers. *Glass Phys. Chem.* **2005**, *31*, 86–95. [CrossRef]

46. Goldstein, R. Electro-optic devices in review. *Lasers Appl.* **1986**, *5*, 67–73.

crystals

MDPI

Article

Comparative Investigation on the Emission Properties of RAl₃(BO₃)₄ (R = Pr, Eu, Tb, Dy, Tm, Yb) Crystals with the Huntite Structure

Enrico Cavalli [1,*] and **Nikolay I. Leonyuk** [2]

[1] Department of Chemical Sciences, Life and Environmental Sustainability, Parma University, 43124 Parma, Italy
[2] Department of Crystallography and Crystal Chemistry, Lomosonov Moscow State University, Moscow 119991, Russia; leon@geol.msu.ru
* Correspondence: enrico.cavalli@unipr.it; Tel.: +39-0521-905436

Received: 10 December 2018; Accepted: 13 January 2019; Published: 16 January 2019

Abstract: The luminescence properties of RAl₃(BO₃)₄ (RAB, with R = Pr, Eu, Tb, Dy, Tm, Yb) huntite crystals grown from $K_2Mo_3O_{10}$ flux were systematically characterized in order to investigate their excitation dynamics, with particular reference to the concentration quenching that in these systems is incomplete. To this purpose, selected excitation, emission, and decay profile measurements on diluted R:YAB crystals were carried out and compared with those of the concentrated compounds. The effects of the energy transfer processes and of the lattice defects, as well as the ion-lattice interactions, have been taken into consideration in order to account for the experimental results.

Keywords: borate crystals; luminescence; rare earth spectroscopy

1. Introduction

Borate crystals of the huntite family with the general formula RX₃(BO₃)₄ (R = lanthanide ions, X = Al, Ga, Fe, Cr) have been the subject of both fundamental and technologically oriented studies [1,2] for more than fifty years. The ability to grow good optical quality single crystals at relatively low temperatures [3], the insensitivity of the huntite structure to the cation replacement, and the presence of a single lanthanide site with defined local symmetry make these materials very suitable for investigations of the structure of the energy levels and the de-excitation mechanisms of the active ions [4,5]. Furthermore, the combination of favorable chemical and physical characteristics like stability, hardness, high UV transparency, and nonlinear optical properties make them attractive for application in several fields: lasers [6,7], scintillators [8], phosphors [9,10] and so on. In general, these crystals require active media, usually constituted by a transparent host matrix, like YAl₃(BO₃)₄ or YGa₃(BO₃)₄, containing luminescent ions in low amounts. This not only for cost reduction purposes, but also because high concentrations of active ions can result in emission quenching effects, a consequence of energy transfer processes whose efficiency depends on the ion nature and on the characteristics of the host. For these reasons, research activity has mainly focused on doped materials (YAl₃(BO₃)₄:R³⁺, hereafter R:YAB, is the most popular) and less attention has been dedicated to the study of the spectroscopic properties of concentrated compounds. Nevertheless, the limited number of investigations carried out on these materials has provided interesting information concerning the effectiveness and the mechanisms of the excitation transfer [11,12], the effect of the rare earth substitution on the structural properties [13], and the development of microchip lasers [14,15]. Consequently, we felt it would be interesting to revisit part of the existing literature and to extend it to unexplored members of the RAl₃(BO₃)₄ (RAB) family, in order to provide a general picture of their emission properties and of the effects of concentration and structure on their luminescence performances.

2. Materials and Methods

2.1. Crystal Growth and Structural Properties

The RAB and R:YAB crystals (R = Pr, Eu, Tb, Dy, Tm, Yb) were grown from $K_2Mo_3O_{10}$–based flux melts in the 1150–900 °C temperature range. The details of the growth procedure are well described in several papers [3,5,13]. The flux growth technique, adopted because the RAB compounds melt incongruently, entails the unavoidable contamination of the crystals by flux components. Investigations in this area have demonstrated that only Mo ions in the tri-, penta- or hexavalent oxidation state are present at a level of some relevance (0.1–0.5%) [1]. The crystals used for the spectroscopic experiments are in the form of rods up to $2 \times 1 \times 1$ mm^3 in size and free from inclusions. Based on X-ray powder diffraction studies [16], the first twelve $RM_3(BO_3)_4$ borate crystals synthesized by Ballman in 1962 [17] were classified structurally as part of the huntite family, $CaMg_3(CO_3)_4$, $R32$ space group (Table 1).

Table 1. Crystallographic characteristics of $RAl_3(BO_3)_4$ (space group $R32$) [1].

Borate	a (Å)	c (Å)
$YAl_3(BO_3)_4$	9.288(3)	7.226(2)
$PrAl_3(BO_3)_4$	9.357(3)	7.312(3)
$EuAl_3(BO_3)_4$	9.319(3)	7.273(3)
$TbAl_3(BO_3)_4$	9.297(3)	7.254(3)
$DyAl_3(BO_3)_4$	9.300(3)	7.249(3)
$TmAl_3(BO_3)_4$	9.282(3)	7.218(3)
$YbAl_3(BO_3)_4$	9.278(3)	7.213(3)

Unlike the carbonate mineral structure, which has only trigonal modification (space group $R32$), the rare-earth-aluminum borates containing large R-cations are represented by three polytypic modifications with space groups $R32$, $C2/c$ and $C2$ (Table 2).

Table 2. Unit cell parameters of monoclinic $RAl_3(BO_3)_4$ modifications.

Borate	a (Å)	b (Å)	c (Å)	β, degree	Sp. Gr.	Ref.
$PrAl_3(BO_3)_4$	7.272(2)	9.362(2)	11.145(3)	103.49	$C2/c$	[18]
$EuAl_3(BO_3)_4$	7.270(4)	9.328(6)	11.074(4)	103.17	$C2/c$	[18]
$EuAl_3(BO_3)_4$	7.230(2)	9.322(4)	16.211(4)	90.72(2)	$C2$	[19]
$TbAl_3(BO_3)_4$	7.220(3)	9.312(4)	11.072(4)	103.20(3)	$C2/c$	[19]

In previous studies [19,20], these polytypes are described in terms of the OD-theory. In this case, monoclinic polytypes are represented in accordance with the transformation of rhombohedral R-cell to monoclinic C-cell with $\beta = 113°$ for both $C2/c$ and $C2$ space groups. The crystals investigated in this work belong to the $R32$ polytype. In the hexagonal huntite lattice, the R^{3+} sites have six-fold oxygen coordination and trigonal prismatic geometry with D_3 point symmetry. The Al^{3+} ions occupy octahedral sites and the B^{3+} ions are surrounded by three oxygen atoms with triangular geometry (Figure 1a).

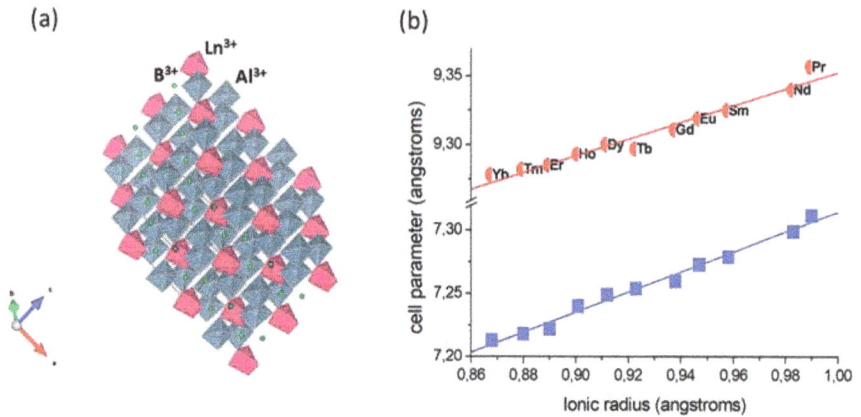

Figure 1. (a) LnAB crystal structure (elaborated using the VESTA software [21]); (b) variation of the cell parameters (taken from Reference [1]) as a function of the Ln^{3+} ionic radii (from Reference [22]).

The Ln^{3+} sites are well separated from one another, the R^{3+}–R^{3+} minimum distance being of the order of 5.8–5.9 Å. This limits energy transfer and concentration quenching processes. It is interesting to note that the cell parameters increase linearly with the ionic radius of the rare earth ions. This dependence can be formalized by the following equations:

$$a\left(\text{Å}\right) = 8.75 + 0.61 \cdot r\left(\text{Å}\right) \tag{1}$$

$$c\left(\text{Å}\right) = 6.52 + 0.79 \cdot r\left(\text{Å}\right) \tag{2}$$

It would be interesting to verify if this model could be extended to other members of the huntite family. Despite their phenomenological nature, they can be useful in different circumstances, like in predicting lattice parameters of unknown compositions, estimating thermodynamic properties, testing data, etc. [23].

2.2. Spectroscopic Measurements

The emission spectra and the decay profiles were measured at room temperature using an Edinburgh FLS1000 (Edinburgh Instruments, Livingston, UK) or a Jobin-Yvon FluoroMax 3 spectrofluorimeter (Horiba, Kyoto, Japan).

3. Results

3.1. PrAB

The emission properties of PrAB were investigated by Koporulina [24] and Malyukin [12]. They observed two band systems centered at 610 and 650 nm, ascribed to the $^1D_2 \rightarrow {}^3H_4$ and $^3P_0 \rightarrow {}^3H_6$ transitions, respectively. An accurate inspection of the spectra shown in Figure 2a allows the assignment of some additional transitions.

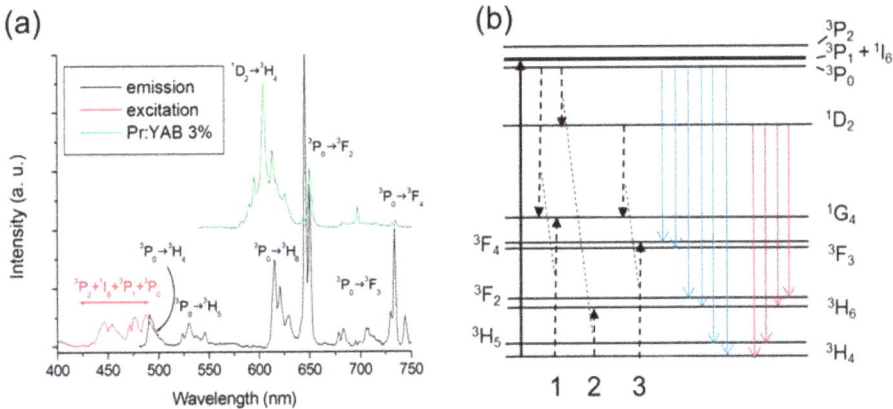

Figure 2. (a) Excitation (monitored wavelength: 613 nm) and emission (excitation at 450 nm) spectrum of PrAB and of Pr^{3+}:YAB. (b) Energy levels scheme and de-excitation mechanisms.

The excitation spectra were assigned to the transition from the $^3H_4$ ground state of Pr^{3+} to the excitation levels indicated in Figure 2a. The spectral components were significantly broadened (FWHM~35–40 cm^{-1} for the most intense transitions) and the comparison with the emission of the diluted compound evidenced the complete quenching of the emission from the $^1D_2$ level, usually predominant in the spectra of the diluted materials [25]. These features are both related to the high content of Pr^{3+} ions. The quenching of the $^1D_2$ emission can be ascribed to a cross-relaxation mechanism, $^1D_2, ^3H_4 \rightarrow ^1G_4, ^3F_4$ (process one in Figure 2b), which is resonant and then effective. It is even because the $^1D_2$ level is efficiently populated through multi-phonon relaxation from the $^3P_0$ one, with the gap between the two levels being of the order of 3500 cm^{-1} and then bridgeable by 3–4 high energy phonons (the phonon cut-off of YAB is 1070 cm^{-1} [26]), and also through cross-relaxation (process two shown in Figure 2b). In addition, the cross-relaxation mechanism depopulating the $^3P_0$ state ($^3P_0, ^3H_4 \rightarrow ^1G_4, ^1G_4$, process three in Figure 2b) was not resonant and less efficient. The combination of these effects meant that only the $^3P_0$ emission was observed in the spectrum of the concentrated crystal.

3.2. EuAB

The excitation and emission spectra of EuAB are reported in Figure 3a. They were consistent with the spectra of the diluted Eu:YAB [9] and Eu:GAB [27] and were assigned accordingly. The excitation spectra were assigned to the transitions from the $^7F_0$ ground state of Eu^{3+} to the excited levels indicated in Figure 3a. The spectral features were relatively narrow (FWHM~20 cm^{-1} for the most intense transitions) indicating that the ion-ion interactions were relatively limited. The EuAB spectra were measured by Kellendonk et al. [11] at liquid helium temperature, in order to demonstrate the presence of Eu^{3+} ions in non-regular sites. In their investigation concerning the concentration quenching of the emission, these authors individuated three possible non-radiative mechanisms depleting the $^5D_0$ emitting level: diffusion-limited migration within the regular Eu^{3+} system, energy transfer between ions in regular and non-regular crystallographic sites, migration to quenching centers and transfer to Mo^{3+} ions present as unwanted impurities. The presence of Eu^{3+} in non-regular sites was consistent with the observation of some extra features in the excitation spectra, revealed through comparison with the excitation spectra of the diluted material (see Figure 4). The decay profile of the luminescence, shown in the inset of Figure 3a, is not exponential and can be well reproduced by a two-exponential function with time constants of 112 and 304 μs, whereas the fit based on the diffusive model of Yokota-Tanimoto [28] does not work.

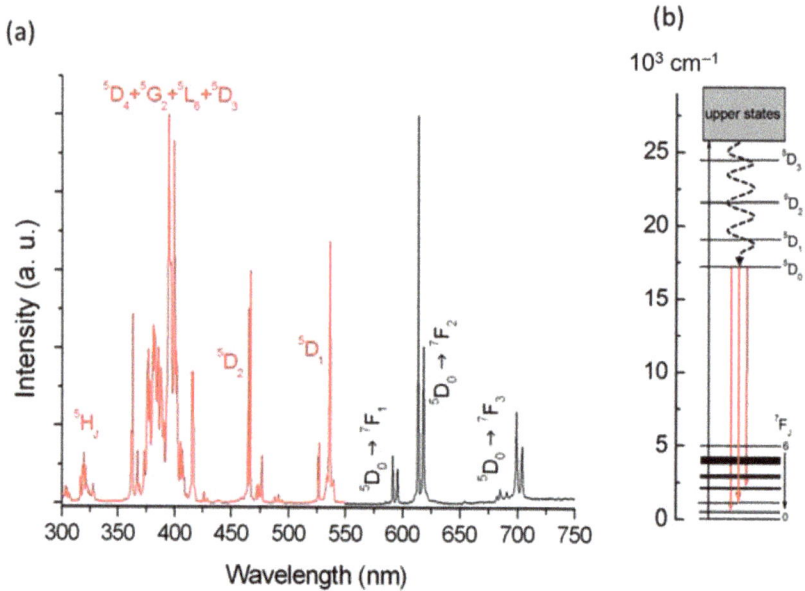

Figure 3. (**a**) Excitation (monitored wavelength: 613 nm) and emission (excitation at 394 nm) spectrum of EuAB. (**b**) Energy levels scheme and de-excitation mechanisms.

This behavior is consistent with the presence of different non-radiative processes, resulting in an overall decrease of the quantum yield to about 22% (estimated according to the ratio of the decay times of the concentrated and diluted crystal), with the radiative lifetime of the $^5D_0$ emitting level being 1.35 ms, as shown in Figure 4.

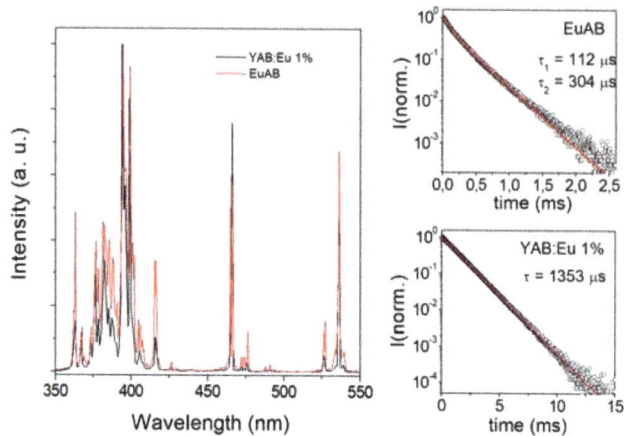

Figure 4. Comparison between the excitation spectra of Eu^{3+}:YAB and EuAB, and decay profiles of their luminescence.

This value was longer than that reported by Kellendonk (1.12 ms) [11]. Considering that EuAB is a fully concentrated material, its efficiency can be considered relatively high.

3.3. TbAB

Apart from the decay profiles, the excitation and emission spectra of TbAB (Figure 5a) were practically identical to those of Tb:YAB 3% and were consistent with previous literature [29,30].

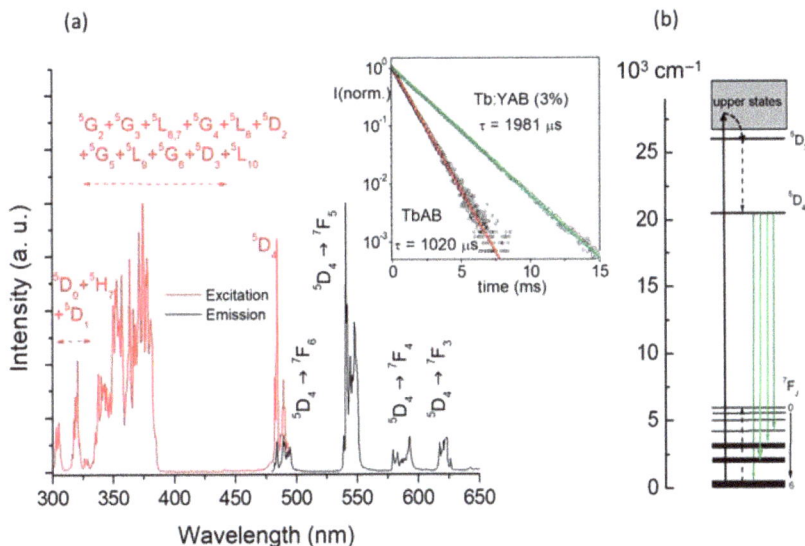

Figure 5. (**a**) Excitation (monitored wavelength: 548 nm) and emission (excitation at 375 nm) spectrum of TbAB. Inset: decay profiles of the Tb:YAB and TbAB emission. (**b**) Energy levels scheme and de-excitation mechanisms.

The observed emission transitions originated from the $^5D_4$ excited state, those from the $^5D_3$ one were quenched through the $^5D_3$, $^7F_6 \rightarrow ^5D_4$, $^7F_0$ cross-relaxation mechanism, as shown in Figure 5b. The excitation spectra were attributed to the transition from the $^7F_6$ ground state of Tb^{3+} to the excited levels indicated in Figure 5a. The temporal profiles of the luminescence were a single exponential and the time constant reduced by less than 50% on passing from the concentrated to the diluted material. Together with the absence of any differences in the structure of the spectra, this means that Tb^{3+} occupies a single site in the huntite lattice, in contrast to Eu^{3+}. In light of the possible energy transfer mechanisms, the comparison between the EuAB and TbAB spectral properties allowed us to infer that the presence of active ions in non-regular or defective sites plays an important role in reducing the efficiency of the material.

3.4. DyAB

To the best of our knowledge, this is the first study to report the excitation and emission spectra, as well as the decay profile, of the luminescence of DyAB (see Figure 6a). The strongest feature was in the yellow region, ascribed to the $^4F_{9/2} \rightarrow ^6H_{13/2}$ transition and interesting for phosphor and laser applications. Similar to the EuAB case, the emission spectrum did not change on passing from the diluted to the fully concentrated compound, whereas the excitation one evidenced a significant broadening and the presence of some extra lines. Thus, it is reasonable to suppose that, in this case, a small part of the doping ions also lie in non-regular sites. It is also important to note that the emitting level can be depleted non-radiatively through a cross-relaxation process ($^4F_{9/2}$, $^6H_{15/2} \rightarrow (^6F_{3/2}$, $^6F_{1/2})$, ($^6H_{9/2}$, $^6F_{11/2})$), shown in Figure 6b. This nearly resonant mechanism accounted for the fast decay of the DyAB emission, whose time constant (2.2 µs) was much shorter than that of the 3% doped Dy:YAB (548 µs).

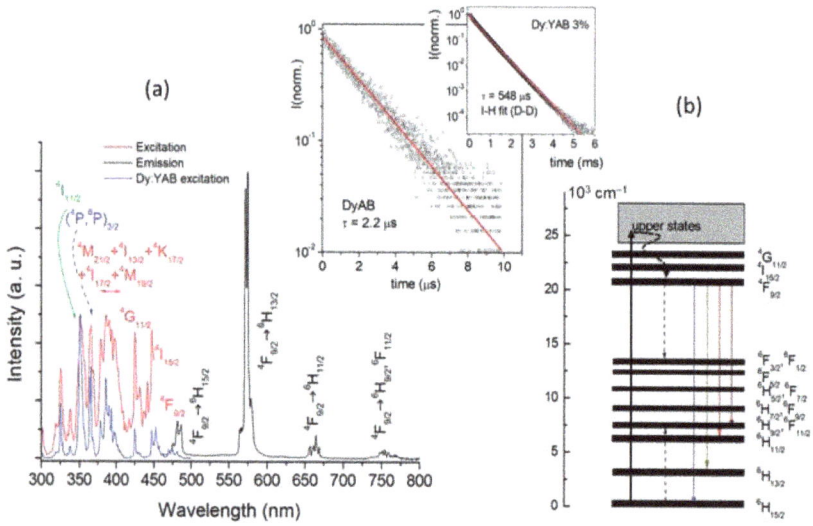

Figure 6. (a) Excitation (monitored wavelength: 575 nm) and emission (excitation at 351 nm) spectrum of DyAB. Inset: decay profiles of the Dy:YAB and DyAB emission. (b) Energy levels scheme and de-excitation mechanisms.

3.5. TmAB

The excitation and emission spectra of TmAB are reported in Figure 7a.

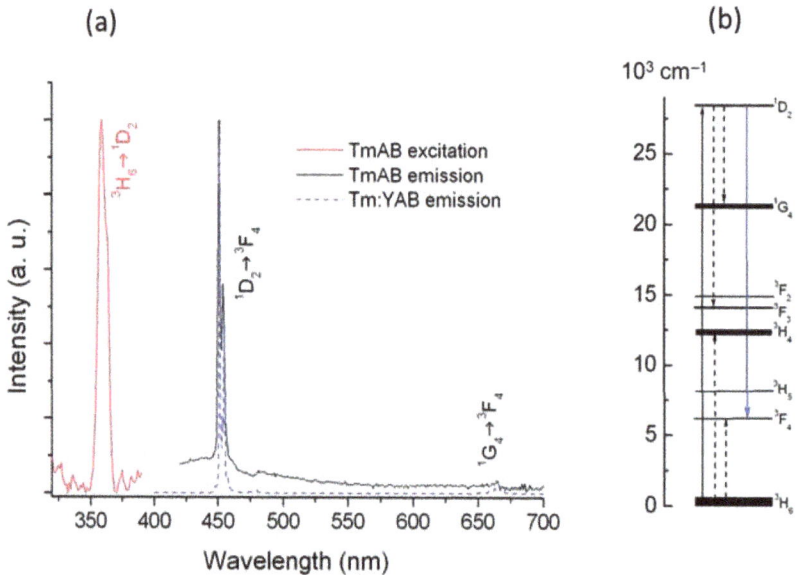

Figure 7. (a) Excitation (monitored wavelength: 450 nm) and emission (excitation at 358 nm) spectrum of TmAB. (b) Energy levels scheme and de-excitation mechanisms.

The luminescence spectrum was measured upon direct excitation into the $^1D_2$ emitting level. It is largely in agreement with the findings reported by Malakhovskii et al. [31], who, however, did not perform excitation or decay time measurements. The corresponding spectrum of Tm:YAB (3%), is shown for the sake of comparison. The maximum intensity of both spectra is normalized to one. The emission spectrum presented a relatively strong band in the blue region, assigned to the $^1D_2 \rightarrow {}^3F_4$ transition, with other features being only barely appreciable at 480 nm ($^1D_2 \rightarrow {}^3H_5$, nearly absent) and at 665 nm ($^1G_4 \rightarrow {}^3F_4$, weaker in the TmAB spectrum). The blue transition overlapped a broad band whose origin is unknown, but is probably due to impurities (Ce^{3+}?). With respect to the emission of the diluted compound, aside from being weaker, it was also relatively broader. The decay profiles were strongly non-exponential in both the diluted and concentrated samples, with the average decay times being 11 µs in the former case and 5 ns in the latter. Considering that the radiative decay time of the $^1D_2$ level is 71 µs [32], it can be concluded that efficient non-radiative processes contributed to depleting the emitting levels. In addition to the energy migration presumably active in TmAB, nearly resonant cross-relaxation processes also occur, as shown in Figure 7b. As a consequence, the concentration quenching, even if incomplete, is rather strong in this material.

3.6. YbAB

The optical spectra of Yb^{3+}-doped YAB have been extensively investigated [33,34] because of the attractiveness of this material for solid-state laser applications. The spectra, shown in Figure 8a for comparison, were in agreement with previous results. As the intensity of the YbAB emission is rather low, it has been amplified in Figure 8 for the sake of comparison. The spectrum was consistent with that published by Popova et al. [35]. However, the decay profile is reported here for the first time. The observed transitions were broadened mainly because of the strong electron-phonon coupling typical of the Yb^{3+}-doped materials. However, the structure of the spectra was different from that of the diluted crystal. In particular, the relative intensity of the excitation and emission bands in the 950–1025 nm range, namely in the vicinity of the 0-0 line at 10188 cm^{-1} (982 nm), was much lower. This is mainly a consequence of reabsorption effects, involving Yb^{3+} ions lying in regular and non-regular sites. In fact, it is known that the optical spectra of Yb^{3+} in YAB are significantly affected by the presence of active ions replacing Al^{3+} or located near to impurities like Mo^{3+} [34–36].

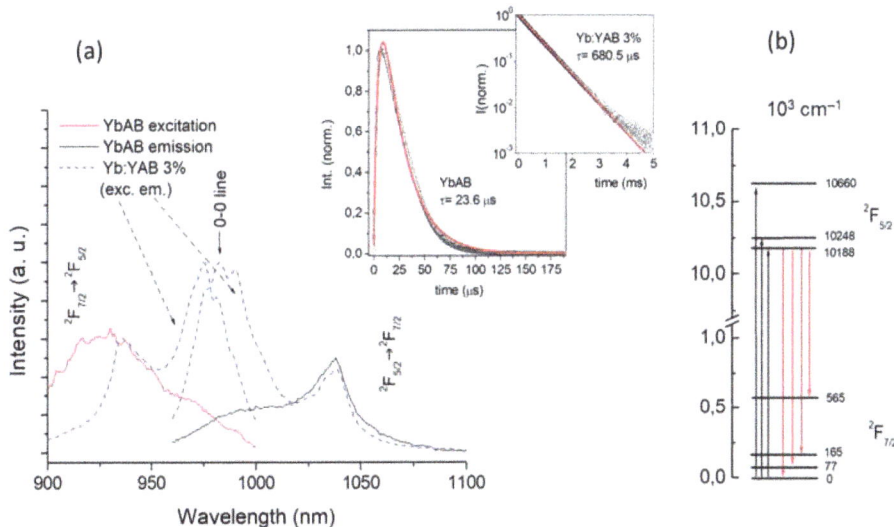

Figure 8. (a) Excitation (monitored wavelength: 1038 nm) and emission (excitation at 940 nm) spectrum of YbAB. Inset: decay profiles of the Yb:YAB and YbAB emission. (b) Energy levels scheme and electronic transitions.

In YbAB, the concentration of these defective species was certainly much higher, as were their effects on the optical features. The build-up in the emission temporal profile and the very short decay constant (23.6 μs versus 680.5 μs for the diluted crystal, see inset of Figure 8) are consistent with the occurrence of energy transfer and migration processes involving active ions located in different environments.

4. Discussion and Conclusions

The emission properties of several members of the RAB (R = lanthanide ion) family were investigated. In these materials, the concentration quenching of the luminescence was incomplete, due to the fact that the crystal sites occupied by the active ions were well separated from one another and the energy transfer processes responsible for the quenching were significantly limited. This finding is in agreement with previous studies [11]. The comparison with the spectra of diluted R^{3+}:YAB has provided information about the quenching degree and the factors responsible for the creation of non-radiative de-excitation pathways. Efficient cross-relaxation channels reduce the emission efficiency of PrAB and TmAB by more than 90%, and their spectra present features ascribable to defect centers and of difficult attribution. The spectra of EuAB and DyAB are rather similar to those of the diluted crystals, but the efficiency is quite high in the former case (22%) and very low (0.4%) in the latter. The mechanism responsible for the quenching of the EuAB emission is the migration to killer centers, whereas that of the DyAB emission is a cross-relaxation mechanism. The emission of TbAB decreases only by about 50% with respect to the diluted crystal, as a consequence of migration processes. Amongst the studied compounds, this is by far the most efficient. Finally, in the case of YbAB, it must be considered that the absorption and emission processes take place between only two electronic states, which are strongly coupled with the lattice. This implies a significant broadening of the spectral features that favors reabsorption processes. This effect is further enhanced by the presence of Yb^{3+} ions in non-regular lattice sites, whose involvement in migration processes results in the reduction of the emission efficiency to about 3%, with respect that of the diluted material. The above considerations are briefly summarized in Table 3.

Table 3. Summary of the luminescence features of the investigated compounds.

Borate	Emitting Level	Efficiency	Quenching Mechanism
$PrAl_3(BO_3)_4$	$^3P_0$	n. d.	Cross-relaxation
$EuAl_3(BO_3)_4$	$^5D_0$	22%	Migration
$TbAl_3(BO_3)_4$	$^5D_4$	51%	Migration
$DyAl_3(BO_3)_4$	$^4F_{9/2}$	0.4%	Cross-relaxation
$TmAl_3(BO_3)_4$	$^1D_2$	7%	Cross-relaxation, migration
$YbAl_3(BO_3)_4$	$^2F_{5/2}$	4%	Migration, reabsorption

In the instance that no data were available in the literature or in the absence of agreement between the published data, the quantum efficiencies were evaluated using the decay times of the diluted crystal as a rough estimation of the radiative lifetimes. Consequently, they must be considered as indicative only. As a final consideration, it can be concluded that the concentration quenching in this class of materials is higher when the emitting level is non-radiatively depleted through cross-relaxation mechanisms, which mostly involve regular centers. In contrast, the migration, dependent to a larger extent on the presence of lattice defects, plays a comparatively minor role. The fact that the quenching was incomplete in the fully concentrated RAB crystals is a consequence of the relatively long distance between the active centers in the huntite lattice.

A better characterization of the energy transfer processes involved in the quenching mechanisms could be performed through low temperature spectroscopic measurements, whereas growth experiments in different solvents, like $BaO-B_2O_3$ or $Li_2B_4O_7$, could be of help for a more detailed identification of the non-regular active ions. Future work is being planned in these directions.

Author Contributions: Conceptualization, spectroscopic measurements and interpretation of the data, manuscript writing: E.C.; Crystal growth and structural characterization, critical reading and review of the manuscript: N.I.L.

Funding: This research was supported in part (N.I. Leonyuk) by the RFBR grants ## 18-05-01085_a and 18-29-12091_MK.

Conflicts of Interest: The authors declare no conflict of interest.

References

1. Leonyuk, N.I.; Leonyuk, L.I. Growth and characterization of $RM_3(BO_3)_4$ crystals. *Prog. Cryst. Growth Charact. Mater.* **1995**, *31*, 179–278. [CrossRef]

2. Leonyuk, N.I. Recent developments in the growth of $RM_3(BO_3)_4$ crystals for science and modern applications. *Prog. Cryst. Growth Charact. Mater.* **1995**, *31*, 279–312. [CrossRef]

3. Nekrasova, L.V.; Leonyuk, N.I. $YbAl_3(BO_3)_4$ and $YAl_3(BO_3)_4$ crystallization from $K_2Mo_3O_{10}$-based high-temperature solutions: Phase relationships and solubility diagrams. *J. Cryst. Growth* **2008**, *311*, 7–9. [CrossRef]

4. Malakhovskii, A.V.; Edelman, I.S.; Solokov, A.E.; Temerov, V.L.; Gnatchenko, S.L.; Kachur, I.S.; Piryatinskaya, V.G. Low temperature absorption spectra of Tm^{3+} ion in $TmAl_3(BO_3)_4$ crystal. *J. Alloys Compd.* **2008**, *439*, 87–94. [CrossRef]

5. Cavalli, E.; Bovero, E.; Magnani, N.; Ramirez, M.O.; Speghini, A.; Bettinelli, M. Optical spectroscopy and crystal-field analysis of $YAl_3(BO_3)_4$ single crystals doped with dysprosium. *J. Phys. Condens. Matter* **2003**, *15*, 1047–1056. [CrossRef]

6. Jaque, D.; Capmany, J.; Molero, F.; García Solé, J. Blue-light laser source by sum-frequency mixing in $Nd:YAl_3(BO_3)_4$. *Appl. Phys. Lett.* **1998**, *73*, 3659–3661. [CrossRef]

7. Dekker, P.; Davies, J.M.; Piper, J.A.; Liu, Y.; Wang, J. 1.1 W CW self-frequency-doubled diode-pumped $Yb:YAl_3(BO_3)_4$ laser. *Opt. Commun.* **2001**, *195*, 431–436. [CrossRef]

8. Aloui-Lebbou, O.; Goutaudier, C.; Kubota, S.; Dujardin, C.; Cohen-Adad, M.T.; Pédrini, C.; Florian, P.; Massiot, D. Structural and scintillation properties of new Ce^{3+}-doped alumino-borate. *Opt. Mater.* **2001**, *16*, 77–86. [CrossRef]

9. Li, G.; Cao, Q.; Li, Z.; Huang, Y. Luminescence properties of $YAl_3(BO_3)_4$ phosphors doped with Eu^{3+} ions. *J. Rare Earths* **2008**, *26*, 792–794. [CrossRef]

10. Li, X.; Wang, Y. Synthesis of $Gd_{1-x}Tb_xAl_3(BO_3)_4$ ($0.05 \leq x \leq 1$) and its luminescence properties under VUV excitation. *J. Lumin.* **2007**, *122–123*, 1000–1002. [CrossRef]

11. Kellendonk, F.; Blasse, G. Luminescence and energy transfer in $EuAl_3B_4O_{12}$. *J. Chem. Phys.* **1981**, *75*, 561–571. [CrossRef]

12. Malyukin, Y.V.; Zhmurin, P.N.; Borysov, R.S.; Roth, M.; Leonyuk, N.I. Spectroscopic and luminescent characteristics of $PrAl_3(BO_3)_4$ crystals. *Opt. Commun.* **2002**, *201*, 355–361. [CrossRef]

13. Benayas, A.; Jaque, D.; García Solé, J.; Leonyuk, N.I.; Bovero, E.; Cavalli, E.; Bettinelli, M. Effects of neodymium incorporation on the structural and luminescence properties of $YAl_3(BO_3)_4$-$NdAl_3(BO_3)_4$ system. *J. Phys. Condens. Matter* **2007**, *19*, 246204. [CrossRef]

14. Jia, G.; Tu, C.; Li, J.; You, Z.; Zhu, Z.; Wu, B. Crystal Structure, Judd-Ofelt Analysis, and Spectroscopic Assessment of a $TmAl_3(BO_3)_4$ Crystal as a New Potential Diode-Pumped Laser near 1.9 μm. *Inorg. Chem.* **2006**, *45*, 9326–9331. [CrossRef] [PubMed]

15. Lutz, F.; Huber, G. Phosphate and borate crystals for high optical gain. *J. Cryst. Growth* **1981**, *52*, 646–649. [CrossRef]

16. Mills, A.D. Crystallographic data for new rare earth borate compounds, $RX_3(BO_3)_4$. *Inorg. Chem.* **1962**, *1*, 960–961. [CrossRef]

17. Ballman, A.A. A new series of synthetic borates isostructural with the carbonate mineral huntite. *Am. Mineral.* **1962**, *47*, 1380–1383.

18. Belokoneva, E.L.; Leonyuk, N.I.; Pashkova, A.V.; Timchenko, T.I. New modifications of rare earth aluminium borates. *Sov. Phys. Crystallogr.* **1988**, *33*, 765–767.

19. Plachinda, P.A.; Belokoneva, E.L. High temperature synthesis and crystal structure of new representatives of the huntite family. *Cryst. Res. Technol.* **2008**, *43*, 157–165. [CrossRef]

20. Belokoneva, E.L.; Timchenko, T.I. Polytype relationships in borate structures whith general formula $RAl_3(BO_3)_4$, R = Y,Nd,Gd. *Sov. Phys. Crystallogr.* **1983**, *28*, 658–661.

21. Momma, K.; Izumi, F. VESTA: A three-dimensional visualization system for electronic and structural analysis. *J. Appl. Crystallogr.* **2008**, *41*, 653–658. [CrossRef]
22. Shannon, R.D. Revised Effective Ionic Radii and Systematic Studies of Interatomic Distances in Halides and Chalcogenides. *Acta Cryst.* **1976**, *A32*, 751–767. [CrossRef]
23. Langley, R.H.; Studgeon, G.D. Lattice parameters and ionic radii of the oxide and fluoride garnets. *J. Sol. State Chem.* **1979**, *30*, 79–82. [CrossRef]
24. Koporulina, E.V.; Leonyuk, N.I.; Hansen, D.; Bray, K.L. Flux growth and luminescence of Ho:YAl$_3$(BO$_3$)$_4$ and PrAl$_3$(BO$_3$)$_4$ crystals. *J. Cryst. Growth* **1998**, *191*, 767–773. [CrossRef]
25. Bartl, M.H.; Gatterer, K.; Cavalli, E.; Speghini, A.; Bettinelli, M. Growth, optical spectroscopy and crystal field investigation of YAl$_3$(BO$_3$)$_4$ single crystals doped with tripositive praseodymium. *Spectrochim. Acta A Mol. Biomol. Spectrosc.* **2001**, *57*, 1981–1990. [CrossRef]
26. Jaque, D.; Ramirez, M.O.; Bausà, L.E.; Garcia Solè, J.; Cavalli, E.; Bettinelli, M.; Speghini, A. Nd$^{3+}\rightarrow$Yb^{3+} energy transfer in the YAl$_3$(BO$_3$)$_4$ nonlinear laser crystal. *Phys. Rev. B* **2003**, *68*, 035118. [CrossRef]
27. Görller-Walrand, C.G.; Huygen, E.; Binnemans, K.; Fluyt, L. Optical absorption spectra, crystal-field energy levels and intensities of Eu^{3+} in GdAl$_3$(BO$_3$)$_4$. *J. Phys. Condens. Matter* **1994**, *6*, 7797–7812. [CrossRef]
28. Yokota, M.; Tanimoto, O. Effects of Diffusion on Energy Transfer by Resonance. *J. Phys. Soc. Jpn.* **1967**, *22*, 779–784. [CrossRef]
29. Kellendonk, F.; Blasse, G. Luminescence and energy transfer in TbAl$_3$B$_4$O$_{12}$. *J. Phys. Chem. Solids* **1982**, *43*, 481–490. [CrossRef]
30. Colak, S.; Zwicker, W.K. Transition rates of Tb^{3+} in TbP$_5$O$_{14}$, TbLiP$_4$O$_{12}$, and TbAl$_3$(BO$_3$)$_4$: An evaluation for laser applications. *J. Appl. Phys.* **1983**, *54*, 2156–2166. [CrossRef]
31. Malakhovskii, A.V.; Valiev, U.V.; Edelman, I.S.; Sokolov, A.E.; Chesnokov, I.Y.; Gudim, I.A. Magneto-optical activity and luminescence of f-f transitions in trigonal crystal TmAl$_3$(BO$_3$)$_4$. *Opt. Mater.* **2010**, *32*, 1017–1021. [CrossRef]
32. Cavalli, E.; Jaque, D.; Leonyuk, N.I.; Speghini, A.; Bettinelli, M. Optical spectra of Tm^{3+}-doped YAl$_3$(BO$_3$)$_4$ single crystals. *Phys. Stat. Sol.* **2007**, *4*, 809–812. [CrossRef]
33. Wang, P.; Davies, J.M.; Dekker, P.; Knowles, D.S.; Piper, J.A.; Lu, B. Growth and evaluation of ytterbium-doped yttrium aluminum borate as a potential self-doubling laser crystal. *J. Opt. Soc. Am. B* **1999**, *16*, 63–69. [CrossRef]
34. Ramirez, M.O.; Bausá, L.E.; Jaque, D.; Cavalli, E.; Speghini, A.; Bettinelli, M. Spectroscopic study of Yb^{3+} centres in the YAl$_3$(BO$_3$)$_4$ nonlinear laser crystal. *J. Phys. Condens. Matter* **2003**, *15*, 7789–7801. [CrossRef]
35. Popova, M.N.; Boldyrev, K.N.; Petit, P.O.; Viana, B.; Bezmaternykh, L.N. High resolution spectroscopy of YbAl$_3$(BO$_3$)$_4$ stoichiometric nonlinear laser crystals. *J. Phys. Condens. Matter* **2008**, *20*, 455210. [CrossRef]
36. Boldyrev, K.N.; Popova, M.N.; Bettinelli, M.; Temerov, V.L.; Gudim, I.A.; Bezmaternykh, L.N.; Loiseau, P.; Aka, G.; Leonyuk, N.I. Quality of the rare earth aluminum borate crystals for laser applications, probed by high-resolution spectroscopy of the Yb^{3+} ion. *Opt. Mater.* **2012**, *34*, 1885–1889. [CrossRef]

crystals

MDPI

Article

Bismuth-Based Oxyborate Piezoelectric Crystals: Growth and Electro-Elastic Properties

Feifei Chen [1], Xiufeng Cheng [1], Fapeng Yu [1,*], Chunlei Wang [2] and Xian Zhao [1]

[1] State Key Laboratory of Crystal Materials, Advanced Research Center for Optics, Shandong University, Jinan 250100, China; ffchen2013@126.com (F.C.); xfcheng@sdu.edu.cn (X.C.); xianzhao@sdu.edu.cn (X.Z.)

[2] School of Physics, Shandong University, Jinan 250100, China; wangcl@sdu.edu.cn

* Correspondence: fapengyu@sdu.edu.cn

Received: 13 December 2018; Accepted: 2 January 2019; Published: 6 January 2019

Abstract: The non-centrosymmetric bismuth-based oxyborate crystals have been extensively studied for non-linear optical, opto-electric and piezoelectric applications. In this work, single crystal growth and electro-elastic properties of α-BiB$_3$O$_6$ (α-BIBO) and Bi$_2$ZnB$_2$O$_7$ (BZBO) crystals are reported. Centimeter-sized α-BIBO and BZBO crystals were grown by using the Kyropoulos method. High resolution X-ray diffraction tests were performed to assess the crystal quality. The full-width at half-maximum values (FWHM) of the rocking curves were evaluated to be 35.35 arcsec and 47.85 arcsec for α-BIBO and BZBO samples, respectively. Moreover, the electro-elastic properties of α-BIBO and BZBO crystals are discussed and summarized, based on which the radial extensional and the face shear vibration modes were studied for potential acoustic device applications. The radial extensional mode electro-mechanical coupling factors k_p were evaluated and found to be 32.0% and 5.5% for α-BIBO and BZBO crystals, respectively. The optimal crystal cuts with face shear mode were designed and found to be $(YZt)/-53°$ (or $(YZt)/37°$ cut) for α-BIBO crystal, and $(ZXt)/\pm45°$ cut for BZBO crystal, with the largest effective piezoelectric coefficients being in the order of 14.8 pC/N and 8.9 pC/N, respectively.

Keywords: α-BiB$_3$O$_6$; Bi$_2$ZnB$_2$O$_7$; single crystal growth; electro-elastic properties

1. Introduction

Oxyborate crystals are important multi-functional crystal materials with comprehensive performances in non-linear optical (NLO), laser, and piezoelectric fields [1–6]. In recent years, the oxyborate crystals, especially the bismuth-based oxyborate crystals, have been paid a great deal of attention for exploring and designing new optical and piezoelectric devices, due to the abundant B-O groups and the bismuth lone pair structure [7,8].

The bismuth-based oxyborate compounds were first discovered in 1962 from the Bi$_2$O$_3$-B$_2$O$_3$ binary phase diagram, where the Bi$_{24}$B$_2$O$_{39}$, Bi$_4$B$_2$O$_9$, Bi$_3$B$_5$O$_{12}$, BiB$_3$O$_6$, and Bi$_2$B$_8$O$_{15}$ crystals were discovered and confirmed [9–13]. Among these compounds, the monoclinic α-BiB$_3$O$_6$ (α-BIBO) crystal was reported to be a valuable NLO material, showing an excellent second harmonic generation (SHG) effect, comparable to the commercialized KTiOPO$_4$ (KTP) crystal [13,14]. Furthermore, active studies were performed for new NLO crystal materials in the ternary Bi$_2$O$_3$-MO-B$_2$O$_3$ system (MO is metal oxide), and a new non-centrosymmetric crystal Bi$_2$ZnB$_2$O$_7$ (BZBO) with orthorhombic symmetry was obtained [15,16]. To date, the α-BIBO and BZBO crystals have been successfully grown in labs, and their NLO and piezoelectric properties have been discussed [17–21]. However, there remains problems during the single crystal growth, and critical growth parameters have not been optimized yet, especially for the α-BIBO crystal. The electro-elastic properties and the piezoelectric vibration modes of the α-BIBO and BZBO crystals are worth studying further. For example, the radial extensional and face shear vibration modes are useful for exploring new acoustic wave devices based on the lamb waves and guided waves [22–24].

In this paper, the monoclinic α-BIBO and orthorhombic BZBO crystals grown by using the Kyropoulos method with modified thermal profiles, are studied. The electro-elastic properties, including the dielectric, elastic, electromechanical and piezoelectric properties of the α-BIBO and BZBO crystals were compared comprehensively, taking advantage of the impedance method. The radial extensional and face shear vibration modes for α-BIBO and BZBO crystals were evaluated for potential acoustic wave device applications.

2. Experimental Section

2.1. Single Crystal Growth

The α-BIBO and BZBO crystals are all congruent compounds and can be grown by using the Kyropoulos method. Firstly, polycrystalline α-BIBO and BZBO powders were synthesized by traditional solid-state reaction technique. The raw materials were weighted according to their stoichiometric ratio (molar ratio of Bi_2O_3: H_3BO_3 = 1:6 for α-BIBO and Bi_2O_3:ZnO:H_3BO_3 = 1:1:2 for BZBO crystals). To prepare the polycrystalline α-BIBO powders, 1.0 mol% H_3BO_3 in excess was added to the raw materials in order to compensate for the evaporation of the B_2O_3 during the solid-state reaction and crystal growth processes, which was proved to be in favor of growing high quality α-BIBO single crystals. Secondly, the raw materials were ground and dry mixed thoroughly for more than 12 h (YGJ-5KG, no use of ball milling), then they were pressed into column blocks. The blocks were sintered at 400 °C for more than 4 h in order to decompose the H_3BO_3 completely, then charged into a platinum (Pt) crucible (φ = 10 cm, H = 12 cm). The crucible was 70% fulfilled with the sintered polycrystalline blocks and loaded into a programmable controlled furnace, which was progressively heated up to 800 °C above the melting points of the α-BIBO and BZBO compounds. Thirdly, the melts were homogenized with a Pt stirrer for more than 24 h, and then kept for 48 h to ensure the uniformity of the melt in the crucible. Fourthly, the temperature was then slowly decreased down to an appropriate temperature for the seeding and crystal growth process.

The principal component of the crystal growth equipment is a tube furnace with electric stove wire as the heating source. The furnace is designed with a special and stable temperature gradient along the axial direction. The schematic diagram of the crystal growth equipment is shown in Figure 1a. It is known that the thermal profile is crucial for crystal growth. Prior to the single crystal growth, the thermal gradient should be determined. In this work, for evaluation of the thermal gradient, the temperature was measured from the bottom of the crucible to the top, and then from the top back to the bottom. To ensure the accuracy of the measurement, the temperature gradients were measured at different temperature points and for each measurement, the dwelling time was kept for 10–15 min. Figure 1b,c present the optimized temperature gradient distributions in the furnace for α-BIBO and BZBO crystals, respectively.

According to our experiments, we found that the axial temperature gradient is sensitive to the growth of BZBO and α-BIBO crystals, since the growth temperature range of these crystals is quite narrow (~4 °C). A narrow axial temperature gradient of <2 °C/cm in the modified thermal profile is beneficial for growing BZBO and α-BIBO crystals stably, thereby improving the crystal quality. In this study, the desired temperature gradients were controlled to be <1 °C/cm and <2 °C/cm approach to the surface of the α-BIBO and BZBO melts, respectively. To avoid the emergence of poly-crystals, a very low cooling rate (0.1–0.5 °C/day) was adapted during the crystal growth.

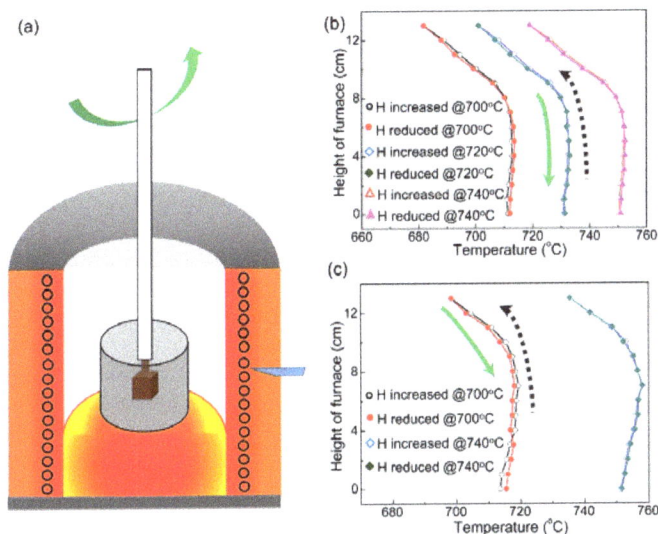

Figure 1. The schematic diagram of the crystal growth furnace (**a**) and the temperature gradient in the furnace used for α-BiB$_3$O$_6$ (α-BIBO) (**b**) and Bi$_2$ZnB$_2$O$_7$ (BZBO) (**c**) crystals.

During the crystal growth process, a bi-directional rotation was adopted to improve the homogeneity of the melt. However, due to the high viscosity, the viscous forces induced by the melt would increase with increasing crystal dimension, which would damage the neck between the crystal seed and bulk crystal [12]. Thus, an appropriate rotation rate should be applied. During the growth of α-BIBO and BZBO crystals, a medium rotation rate of 10–15 rpm was adapted and controlled. The crystal growth period was selected to be 2 to 8 weeks. When the crystal reached the expected dimension, it was pulled out of the melt slowly and hung over the melt for 1–2 h, then cooled down to room temperature with a low rate of 5–20 °C. Figure 2 presents the as-grown α-BIBO and BZBO crystals with distinguished habitual facets. Though some impurities (polycrystalline raw materials) adhered on the surface, the total crystals were transparent inside. The impurities were generated when pulling the crystal out of the melt.

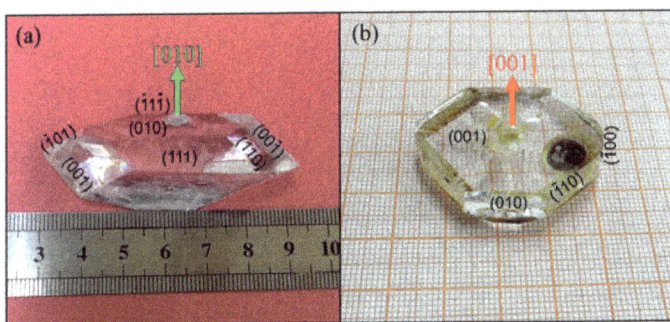

Figure 2. The as-grown α-BIBO (**a**) and BZBO (**b**) single crystals.

2.2. High-Resolution X-Ray Diffraction

The qualities of the grown α-BIBO and BZBO crystals can be evaluated by using the high-resolution X-ray diffraction (HRXRD) method. In our study, the HRXRD tests were performed using a Bruker-axs D5005HR diffractometer (Bruker-axs, Karlsruhe, Germany) equipped with a two-crystal Ge (220) monochromator set for Cu-Kα1 radiation (λ = 1.54056 Å). The accelerating voltage and tube current were 20 kV and 30 mA, and the step time and size were 0.1 s and 0.001°, respectively. The measured crystal wafers were (010) facet for α-BIBO crystals and (001) facet for BZBO crystals, with both sides polished.

2.3. Characterization of Electro-Elastic Properties

The number of independent electro-elastic constants is relevant to the crystal symmetry. It is clear that the α-BIBO belongs to the monoclinic symmetry with point group 2, while the BZBO crystal possesses orthorhombic symmetry with point group mm2. The related lattice parameters for α-BIBO and BZBO crystals are presented in Table 1 [15,25]. Therefore, the electro-elastic constants of α-BIBO and BZBO crystals are very different. The electro-elastic constants of these crystals were characterized by using the impedance method. The related crystal cuts, vibration modes, and equations for evaluating the independent dielectric, elastic, electromechanical, and piezoelectric constants were studied and reported in our previous work [18,21]. For the radial extensional mode, the effective electromechanical coupling factors k_p were evaluated by measuring the Y- and Z-oriented discs for α-BIBO and BZBO crystals, respectively. The disc-shaped samples (6 pieces), with different dimension ratios, were prepared and vacuum-sputtered with platinum films (100–200 nm) on the two parallel faces. The capacitance, resonance and anti-resonance frequencies of the prepared samples were measured and recorded by using the multi-frequency LCR meter (Agilent 4263B) and impedance-phase gain analyzer HP4194A.

Table 1. Crystal lattice parameters for α-BIBO and BZBO crystals [15,25].

Empirical Formula	α-BiB$_3$O$_6$	Bi$_2$ZnB$_2$O$_7$
Formula weight	337.41	616.97
symmetry	monoclinic	orthorhombic
space group	C2	Pba2
a (Å)	7.116 (2)	10.8268 (4)
b (Å)	4.993 (2)	11.0329 (4)
c (Å)	6.508 (3)	4.8848 (2)
V (Å^3)	222.69	583.49 (19)
Z	2	4
density (Mg/m^3)	5.033	7.036

3. Results and Discussion

3.1. Crystal Quality Evaluation

The crystal quality was assessed by using the HRXRD method. Figure 3 gives the rocking curves of the two oxyborate crystals, where the crystal samples for α-BIBO and BZBO crystals were selected and prepared from the parts away from the seeding region. The diffraction peaks were symmetrical and the full-width at half-maximum values (FWHM) were obtained and found to be 35.35 arcsec and 47.85 arcsec for α-BIBO and BZBO crystals, respectively. In order to characterize the homogeneity of the growth crystal, the HRXRD tests for different parts (away from the seeding region and close to the seeding region) of BZBO crystals were performed. The FWHM values for the crystal parts away from the seeding region were found smaller for both the α-BIBO and BZBO crystals. The relative high FWHM value of the BZBO crystal sample prepared from the seeding region indicates that the initial growth process is vital for high quality crystal growth. It was also found that the crystal quality would improve as the crystal grows. In this study, the crystal samples were prepared from the good quality parts to get rid of the possible effects of crystal defects.

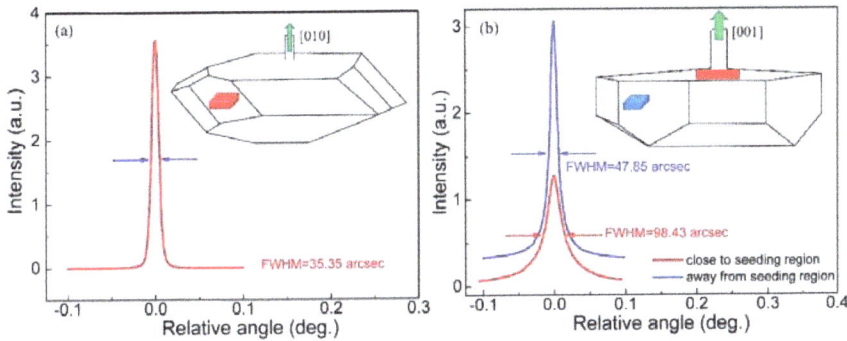

Figure 3. Rocking curves of the (010) wafers for α-BIBO (**a**) and (001) wafers for BZBO (**b**).

3.2. Electro-Elastic Constants

The electro-elastic constants for α-BIBO and BZBO crystals, measured at room temperature, are summarized in Table 2. Compared to the BZBO crystal, the α-BIBO showed stronger piezoelectric response, where the longitudinal piezoelectric charge coefficient d_{22} was found to be 40 pC/N. In contrast, the BZBO crystal exhibited relatively weak piezoelectric properties, the largest piezoelectric charge coefficient was found for d_{32}, being −6.4 pC/N, and the largest elastic compliance was obtained for s_{66}, being 20.5 pm^2/N. The large piezoelectric activity of α-BIBO crystal was presumed to be associated with the large structure distortions and net dipole moments. Figure 4 exhibits the schematic crystal structures for α-BIBO and BZBO crystals, where the α-BIBO crystal consists of one type of Bi-O octahedron, while the BZBO crystal possesses two types of Bi-O octahedra. Different from the α-BIBO crystal, the directions of the dipole moment for the two types of Bi-O octahedra in BZBO crystal were partly offset, which weakened the piezoelectric response in BZBO crystal, thus the effective piezoelectric coefficients for BZBO crystal were lower than the α-BIBO crystal [21,26,27].

Table 2. The electro-elastic constants of α-BIBO and BZBO crystals measured at room temperature [18,21].

Elastic Compliances s_{ij}^E (pm^2·N^{-1})												
s_{11}	s_{12}	s_{13}	s_{15}	s_{22}	s_{23}	s_{25}	s_{33}	s_{35}	s_{44}	s_{46}	s_{55}	s_{66}
BIBO 36.2	−48.0	2.9	17.9	85.0	−2.6	−23.8	10.2	9.3	65.0	11.5	26.5	19.1
BZBO 8.2	−3.9	−1.5	\	11.8	−3.5	\	8.2	\	17.3	\	17.2	20.5
Elastic Stiffnesses c_{ij}^E (10^{10} N·m^{-2})												
c_{11}	c_{12}	c_{13}	c_{15}	c_{22}	c_{23}	c_{25}	c_{33}	c_{35}	c_{44}	c_{46}	c_{55}	c_{66}
BIBO 12.4	6.1	1.0	−3.2	4.7	−0.9	0.4	15.7	−7.0	1.7	−1.0	8.8	5.9
BZBO 17.0	7.4	6.3	\	13.1	6.9	\	16.4	\	5.8	\	5.8	4.9
Relative Dielectric Permittivities $\varepsilon_{ij}^T/\varepsilon_0$												
ε_{11}	ε_{13}	ε_{22}	ε_{33}									
BIBO 12.0	−1.4	8.4	13.8									
BZBO 36.8	\	18.5	18.3									
Piezoelectric Charge Coefficients d_{ij} (pC/N)												
d_{14}	d_{15}	d_{16}	d_{21}	d_{22}	d_{23}	d_{24}	d_{25}	d_{31}	d_{32}	d_{33}	d_{34}	d_{36}
BIBO 10.9	\	13.9	16.7	40.0	2.5	\	4.3	\	\	\	18.7	13.0
BZBO \	1.4	\	\	\	\	−5.5	\	2.5	−6.4	1.1	\	\
Electromechanical Coupling Factors k_{ij} (%)												
k_{14}	k_{15}	k_{16}	k_{21}	k_{22}	k_{23}	k_{24}	k_{25}	k_{31}	k_{32}	k_{33}	k_{34}	k_{36}
BIBO 13.1	\	30.9	32.1	50.0	9.2	\	9.6	\	\	\	21.0	26.9
BZBO \	1.8	\	\	\	\	10.7	\	8.8	14.5	3.1	\	\

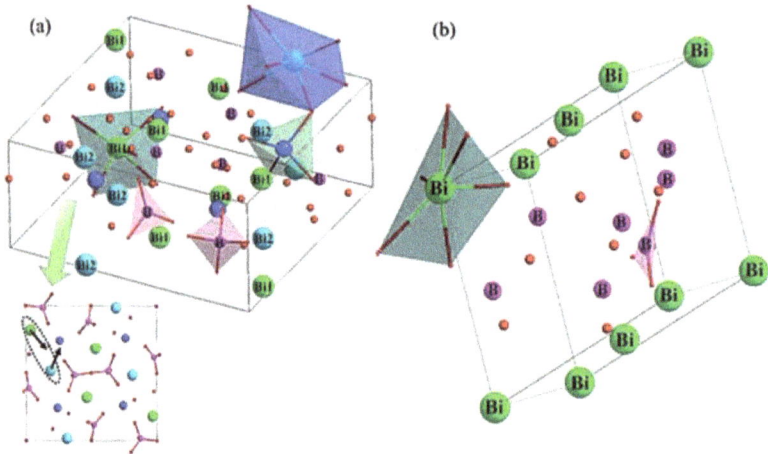

Figure 4. Crystal structures of BZBO (**a**) and α-BIBO (**b**) crystals.

3.3. Characterization of Radial Extensional Vibration Mode

For the acoustic wave devices, there are many kinds of acoustic modes that can be utilized in a very wide frequency range (kHz~GHz). Among these acoustic waves, the lamb wave has been extensively studied [28–30]. The lamb wave can be excited using the radial extensional vibration mode. For the α-BIBO and BZBO crystals, the electromechanical coupling factor k_p, relevant to the radial extensional vibration mode, was evaluated experimentally. The disc-shaped samples with different ratios were prepared for determining the k_p values. Using the empirical equation [31], as well as the measured resonant and anti-resonant frequencies, the radial extensional vibration mode electromechanical coupling factors k_p, for α-BIBO and BZBO crystals, were obtained.

Figure 5 shows the obtained k_p values for the Y-oriented α-BIBO and Z-oriented BZBO crystal discs. The small inset gives the recorded impedance-frequency spectra for the radial extensional vibration modes. It was observed that the difference of resonant and anti-resonant frequencies for the radial extensional mode of α-BIBO was larger than that of BZBO crystal. The k_p values were determined to be in the order of 32.0% and 5.5% for the α-BIBO and BZBO crystals, respectively.

3.4. Characterization of the Face Shear Vibration Mode

The face shear vibration mode could excite the guided wave along the crystal dimensional orientation, which could also be used for sensor application. According to the IEEE standard on piezoelectricity, there are three possible independent face shear mode piezoelectric coefficients in crystal, i.e., d_{14}, d_{25}, and d_{36}. In this work, the face shear vibration modes of α-BIBO and BZBO crystals were discussed, taking advantage of the anisotropy of crystal materials. In order to design the optimum crystal cuts with large piezoelectric coefficients, the orientation dependences of the d_{14}, d_{25}, and d_{36} corresponding to the XY, YZ, and ZX crystal cuts respectively were investigated (rotation angle around physical X-, Y-, and Z-axes was varied from $-90°$ to $90°$). Results are given in Figures 6 and 7.

For α-BIBO single crystal, as shown in Figure 6, the shear mode piezoelectric coefficient d_{25} was found to be more sensitive to the rotation angle, compared with d_{14} and d_{36}. The variations of the d_{14} and d_{36}, rotated along the three physical axes, were symmetrically distributed, referring to the rotation angle at $0°$. The piezoelectric coefficient d_{25} changed from -13 pC/N to 4.3 pC/N, and from -10.9 pC/N to 4.3 pC/N, as the angle rotated around the X- and Z-axes, respectively. In contrast, the largest piezoelectric coefficient d_{25} (YZ cut) was obtained and found to be in the order of 14.8 pC/N and -14.8 pC/N, when the rotation angle reached at $37°$ and $-53°$ around the Y-axis (the small inset of Figure 6), respectively, more than two times that of langasite crystal ($d_{14} = -6.01$ pC/N) [32].

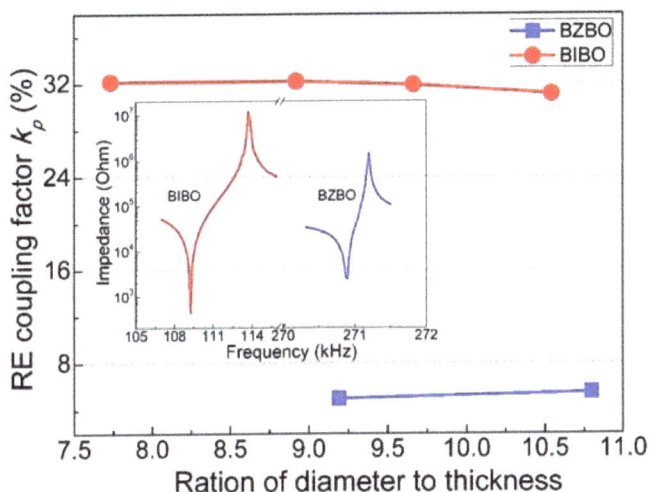

Figure 5. The radial extensional electromechanical coupling factor k_p for the α-BIBO and BZBO crystals.

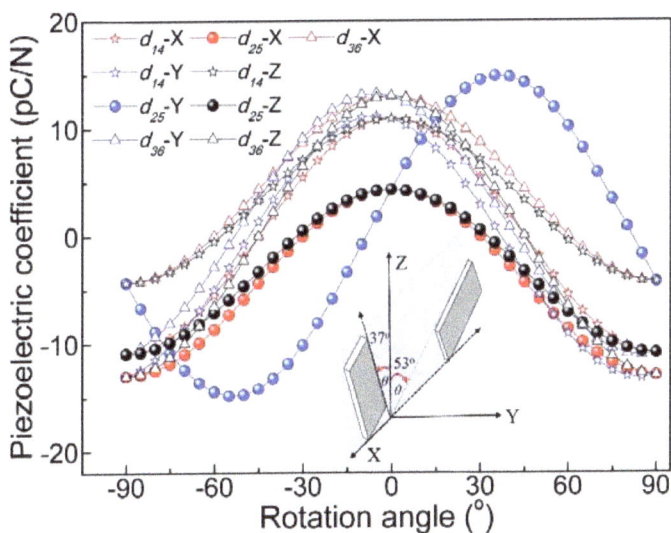

Figure 6. Orientation dependence of piezoelectric coefficient rotated around X-, Y- and Z-axes for α-BIBO crystals. The small inset shows the crystal cuts ($YZt/37°$ and $YZt/-53°$) with largest d_{25} value.

The face shear mode piezoelectric coefficients d_{14}, d_{25} and d_{36} for BZBO crystal are also studied and the results are given in Figure 7. The BZBO crystal was found to show very limited face shear modes, due to the crystal symmetry. The effective face shear mode piezoelectric coefficients d_{14}, d_{25} and d_{36} were only observed when rotated around the physical Z-axis. The effective piezoelectric coefficient d_{14} equaled to the d_{25}, as a function of rotation angle around the Z-axis. The largest values were found to be ±3.5 pC/N when the rotation angles reached $\pm45°$. Differently, the piezoelectric coefficient d_{36} varied from -8.9 pC/N to 8.9 pC/N, when the rotation angle shifted from $-90°$ to $90°$. The maximum value of d_{36} (ZX crystal cut) was determined to be in the order of ±8.9 pC/N when the rotation angle approached to $\pm45°$. Thus, the optimal face shear mode piezoelectric crystal cuts were found to be (ZXt)/$\pm45°$ for BZBO crystal, as presented in the small inset of Figure 7.

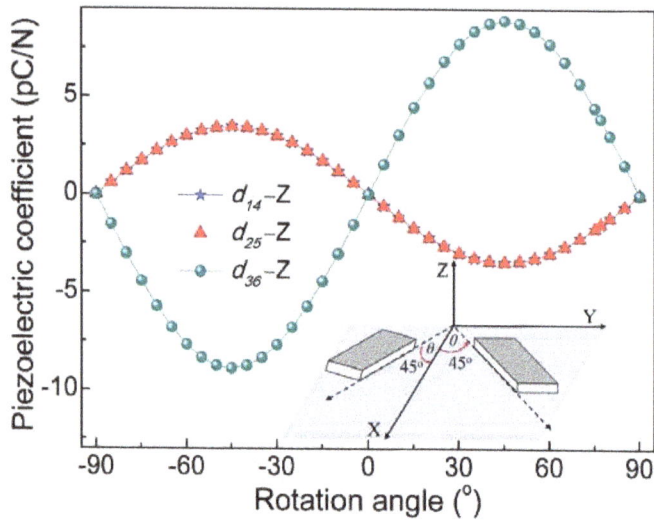

Figure 7. Orientation dependences of piezoelectric coefficient rotated around Z-axis for BZBO crystals. The small inset shows the optimum crystal cuts ((ZXt)/±45° crystal cuts) with the largest d_{36}.

4. Conclusions

In this study, single crystal growth of bismuth-based oxyborate crystals α-BIBO and BZBO was introduced. The quality of the α-BIBO and BZBO crystals grown by the Kyropoulos method was assessed using the HRXRD method. The results indicate that the crystals part aways from the seeding region exhibiting high crystal quality. The electro-elastic properties of α-BIBO and BZBO crystals were evaluated at room temperature taking advantage of the impedance method, where the α-BIBO crystal showed stronger piezoelectric response than the BZBO crystal. In addition, the radial extensional and face shear vibration modes were studied. The radial extensional electromechanical coupling factors k_p were evaluated to be 32.0% and 5.5% for α-BIBO and BZBO crystals, respectively. Furthermore, the optimal face shear mode crystal cuts were obtained and found to be (YZt)/−53° and (YZt)/37° for α-BIBO crystal, and (ZXt)/±45° cuts for BZBO crystal. These crystal cuts are a potential for acoustic wave device application.

Author Contributions: Conceptualization, F.C. and F.Y.; Investigation, F.C.; Methodology, F.C., X.C., F.Y. and C.W.; Resources, F.Y. and X.Z.; Validation, F.Y. and X.Z; Writing—original draft, F.C.; Writing—review and editing, F.Y.

Funding: This research was financially supported by the Primary Research and Development Plan of Shandong Province (2017CXGC0413) and the National Nature Science Foundation of China (51872165).

Acknowledgments: The authors would like to thank Fangming Zuo at Jinan SCK photonics Co., Ltd. for sample preparation.

Conflicts of Interest: The authors declare no conflicts of interest.

References

1. Yang, Y.; Jiang, X.X.; Lin, Z.S.; Wu, Y.C. Borate-based ultraviolet and deep-ultraviolet nonlinear optical crystals. *Crystals* **2017**, *7*, 95. [CrossRef]
2. Bubnova, R.; Volkov, S.; Albert, B.; Filatov, S. Borates-crystal structures of prospective nonlinear optical materials: High anisotropy of the thermal expansion caused by anharmonic atomic vibrations. *Crystals* **2017**, *7*, 93. [CrossRef]
3. Wu, Y.C.; Sasaki, T.; Nakai, S.; Yokotani, A.; Tang, H.G.; Chen, C.T. CsB₃O₅: A new nonlinear optical crystal. *Appl. Phys. Lett.* **1993**, *62*, 2614–2615. [CrossRef]

4. Aka, G.; Kahn-Harari, F.; Mougel, F.; Vivien, D. Linear- and nonlinear-optical properties of a new gadolinium calcium oxoborate crystal, $Ca_4GdO(BO_3)_3$. *J. Opt. Soc. Am. B* **1997**, *14*, 2238–2247. [CrossRef]
5. Sasaki, T.; Mori, Y.; Yoshimura, M.; Yap, Y.K.; Kamimura, T. Recent development of nonlinear optical borate crystals: Key materials for generation of visible and UV light. *Mater. Sci. Eng.* **2000**, *30*, 1–54. [CrossRef]
6. Mori, Y.; Yap, Y.K.; Kamimura, T.; Yoshimura, M.; Sasaki, T. Recent development of nonlinear optical borate crystals for UV generation. *Opt. Mater.* **2002**, *19*, 1–5. [CrossRef]
7. Hu, C.; Mutailipu, M.; Wang, Y.; Guo, F.J.; Yang, Z.H.; Pan, S.L. The activity of lone pair contributing to SHG response in bismuth borates: A combination investigation from experiment and DFT calculation. *Phys. Chem. Chem. Phys.* **2017**, *19*, 25270–25276. [CrossRef]
8. Lin, Z.S.; Wang, Z.Z.; Chen, C.T.; Lee, M.H. Mechanism for linear and nonlinear optical effects in monoclinic bismuth borate (BiB_3O_6) crystal. *J. Appl. Phys.* **2001**, *90*, 5585–5590. [CrossRef]
9. Levin, E.M.; Mcdaniel, C.L. The System Bi_2O_3-B_2O_3. *J. Am. Ceram. Soc.* **1962**, *45*, 355–360. [CrossRef]
10. Burianer, M.; Mṿhlberg, M. Crystal growth of boron sillenite $Bi_{24}B_2O_{39}$. *Cryst. Res. Technol.* **1997**, *32*, 1023–1027. [CrossRef]
11. Filatov, S.; Shepelev, Y.; Bubnova, R.; Sennova, N.; Egorysheva, A.V.; Kargin, Y.F. The study of $Bi_3B_5O_{12}$: Synthesis, crystal structure and thermal expansion of oxoborate $Bi_3B_5O_{12}$. *J. Solid State Chem.* **2004**, *177*, 515–522. [CrossRef]
12. Becker, P.; Liebertz, J.; Bohatý, L. Top-seeded growth of bismuth triborate, BiB_3O_6. *J. Cryst. Growth* **1999**, *203*, 149–155. [CrossRef]
13. Zhang, K.C.; Chen, X.A.; Wang, X.M. Review of study on bismuth triborate (BiB_3O_6) crystal. *J. Synth. Cryst.* **2005**, *34*, 438–443.
14. Wang, Z.P.; Teng, B.; Fu, K.; Xu, X.G.; Song, R.B.; Du, C.L.; Jiang, H.D.; Wang, J.Y.; Liu, Y.G.; Shao, Z.S. Efficient second harmonic generation of pulsed laser radiation in BiB_3O_6 (BIBO) crystal with different phase matching directions. *Opt. Commun.* **2002**, *202*, 217–220. [CrossRef]
15. Barbier, J.; Penin, N.; Cranswick, L.M. Melilite-type borates $Bi_2ZnB_2O_7$ and $CaBiGaB_2O_7$. *Chem. Mater.* **2005**, *17*, 3130–3136. [CrossRef]
16. Li, F.; Pan, S.L.; Hou, X.L.; Yao, J. A novel nonlinear optical crystal $Bi_2ZnOB_2O_6$. *Cryst. Growth Des.* **2009**, *9*, 4091–4095. [CrossRef]
17. Teng, B.; Wang, J.Y.; Wang, Z.P.; Hu, X.B.; Jiang, H.D.; Liu, H.; Cheng, X.F.; Dong, S.M.; Liu, Y.G.; Shao, Z.S. Crystal growth, thermal and optical performance of BiB_3O_6. *J. Cryst. Growth* **2001**, *233*, 282–286. [CrossRef]
18. Yu, F.P.; Lu, Q.M.; Zhang, S.J.; Wang, H.W.; Cheng, X.F.; Zhao, X. High-performance, high-temperature piezoelectric BiB_3O_6 crystals. *J. Mater. Chem. C* **2015**, *3*, 329–338. [CrossRef]
19. Li, F.; Hou, X.L.; Pan, S.L.; Wang, X. Growth, structure, and optical properties of a congruent melting oxyborate, $Bi_2ZnOB_2O_6$. *Chem. Mater.* **2009**, *21*, 2846–2850. [CrossRef]
20. Chen, F.F.; Wang, X.L.; Wei, L.; Yu, F.P.; Tian, S.W.; Jiang, C.; Li, Y.L.; Cheng, X.F.; Wang, Z.P.; Zhao, X. Thermal properties and CW laser performances of pure and Nd doped $Bi_2ZnB_2O_7$ single crystals. *CrystEngComm* **2018**, *20*, 7094–7099. [CrossRef]
21. Chen, F.F.; Jiang, C.; Tian, S.W.; Yu, F.P.; Cheng, X.F.; Duan, X.L.; Wang, Z.P.; Zhao, X. Electroelastic features of piezoelectric $Bi_2ZnB_2O_7$ crystal. *Cryst. Growth Des.* **2018**, *18*, 3988–3996. [CrossRef]
22. Kauffmann, P.; Ploix, M.A.; Chaix, J.F.; Gueudré, C.; Corneloup, G.; Baqué, F. Study of lamb waves for non-destructive testing behind screens. *EPJ Web Conf.* **2018**, *170*, 1–3. [CrossRef]
23. Rguiti, M.; Grondel, S.; El youbi, F.; Courtois, C.; Lippert, M.; Leriche, A. Optimized piezoelectric sensor for a specific application: Detection of lamb waves. *Sens. Actuators A* **2006**, *126*, 362–368. [CrossRef]
24. Giurgiutiu, V. *Structural Health Monitoring, with Piezoelectric Wafer Active Sensors*; Elsevier: Amsterdam, The Netherlands, 2007.
25. Fröhlich, V.R.; Bohatý, U.L.; Lieberta, J. Die kristallstruktur von wismutborat, BiB_3O_6. *Acta Cryst.* **1984**, *C40*, 343–344. [CrossRef]
26. Brese, N.E.; Keeffe, M.O. Bond-valence parameters for solids. *Acta Cryst.* **1991**, *B47*, 192–197. [CrossRef]
27. Maggard, P.A.; Nault, T.S.; Stern, C.L.; Kenneth, R.P. Alignment of acentric $MoO_3F_3^{3-}$ anions in a polar material:$(Ag_3MoO_3F_3)(Ag_3MoO_4)Cl$. *J. Solid State Chem.* **2003**, *175*, 27–33. [CrossRef]
28. Su, Z.Q.; Ye, L.; Lu, Y. Guided lamb waves for identification of damage in composite structures: A review. *J. Sound Vib.* **2006**, *295*, 753–780. [CrossRef]

29. Kuypers, J.H.; Pisano, A.P. Interpolation technique for fast analysis of surface acoustic wave and lamb wave devices. *Jpn. J. Appl. Phys.* **2009**, *48*, 07GG07. [CrossRef]

30. Schmitt, M.; Olfert, S.; Rautenberg, J.; Lindner, G.; Henning, B.; Reindl, L.M. Multi reflection of lamb wave emission in an acoustic waveguide sensor. *Sensors* **2013**, *13*, 2777–2785. [CrossRef]

31. Li, Y.; Qin, Z.K. *Measurement of Piezoelectric and Ferroelectric Materials*; Science Press: Beijing, China, 1984.

32. Bohm, J.; Chilla, E.; Flannery, C.; Frohlich, H.J.; Hauke, T.; Heimann, R.B.; Hengst, M.; Straube, U. Czochralski growth and characterization of piezoelectric single crystals with langasite structure: $La_3Ga_5SiO_{14}$(LGS), $La_3Ga_{5.5}Nb_{0.5}O_{14}$(LGN) and $La_3Ga_{5.5}Ta_{0.5}O_{14}$ (LGT) II. Piezoelectric and elastic properties. *J. Cryst. Growth* **2000**, *216*, 293–298. [CrossRef]

crystals

MDPI

Article

Twins in $YAl_3(BO_3)_4$ and $K_2Al_2B_2O_7$ Crystals as Revealed by Changes in Optical Activity

Johannes Buchen [1,2] , **Volker Wesemann** [1,*], **Steffen Dehmelt** [1,3], **Andreas Gross** [1] and **Daniel Rytz** [1]

[1]   FEE GmbH, Struthstr. 2, 55743 Idar-Oberstein, Germany; jobuchen@caltech.edu (J.B.);
      sdehmelt@gmx.de (S.D.); gross@fee-io.de (A.G.); rytz@fee-io.de (D.R.)

[2]   Division of Geological and Planetary Sciences, California Institute of Technology, 1200 E. California Blvd.,
      Pasadena, CA 91125, USA

[3]   Hochschule Trier, Umwelt-Campus Birkenfeld, Campusallee, 55768 Hoppstädten-Weiersbach, Germany

*   Correspondence: wesemann@fee-io.de

Received: 7 December 2018 ; Accepted: 20 December 2018; Published: 22 December 2018

Abstract: Many borate crystals feature nonlinear optical properties that allow for efficient frequency conversion of common lasers down into the ultraviolet spectrum. Twinning may degrade crystal quality and affect nonlinear optical properties, in particular if crystals are composed of twin domains with opposing polarities. Here, we use measurements of optical activity to demonstrate the existence of inversion twins within single crystals of $YAl_3(BO_3)_4$ (YAB) and $K_2Al_2B_2O_7$ (KABO). We determine the optical rotatory dispersion of YAB and KABO throughout the visible spectrum using a spectrophotometer with rotatable polarizers. Space-resolved measurements of the optical rotation can be related to the twin structure and give estimates on the extent of twinning. The reported dispersion relations for the rotatory power of YAB and KABO may be used to assess crystal quality and to select twin-free specimens.

Keywords: NLO crystals; frequency conversion; second harmonic generation; YAB; $YAl_3(BO_3)_4$; KABO; $K_2Al_2B_2O_7$; optical activity; optical rotatory dispersion; inversion twin

1. Introduction

A series of borates with the general formula $RAl_3(BO_3)_4$ (R = Y or rare earth element (REE)) adopts the crystal structure of the carbonate mineral huntite, $CaMg_3(CO_3)_4$ [1–3]. In this structure type, the coplanar arrangement of $[BO_3]^{3-}$ groups amplifies the intrinsic nonlinear dielectric susceptibility of the $[BO_3]^{3-}$ group and combines it with strong birefringence and a spectral transparency that extends into the ultraviolet (UV) region [4–6]. $YAl_3(BO_3)_4$ (YAB) is transparent down to about 170 nm [7,8] and shows an acceptable effective nonlinear optical (NLO) coefficient [9]. The refractive indices of YAB permit phase matching for several types of frequency conversion to UV light including second harmonic generation (SHG) from 532 nm to 266 nm [7–9]. In contrast to many other NLO borates, YAB is non-hygroscopic and hard (Mohs hardness 7.5 [2]) as compared to β-BaB_2O_4 (BBO) [10], LiB_3O_5 (LBO) [11], and $CsLiB_6O_{10}$ (CLBO) [12]. Albeit not a huntite borate, the material $K_2Al_2B_2O_7$ (KABO) shares many of the advantages of YAB [13,14]. Both YAB and KABO are free of toxic elements, such as beryllium, an essential component of other proposed NLO materials including $Sr_2Be_2B_2O_7$ (SBBO) [15] and $KBe_2BO_3F_2$ (KBBF) [16,17]. The combinations of suitable optical, chemical, and mechanical properties turn YAB and KABO into promising crystals for NLO applications in the UV spectral region.

Both pure and REE-doped YAB crystals, however, typically show a lamellar microstructure as revealed by X-ray diffraction topography [18], surface etching patterns [19,20], and atomic force microscopy [21]. A similar microstructure has been detected in crystals of the borate huntite $Y_{0.57}La_{0.72}Sc_{2.71}(BO_3)_4$ (YLSB) [22,23]. Based on these observations, the lamellae have been interpreted as polysynthetic inversion twins with reflection on $\{11\bar{2}0\}$ planes as the twin operation and adjacent twin domains sharing planes parallel to $\{10\bar{1}1\}$ [18,21,24]. This twin law, i.e., reflection on $\{11\bar{2}0\}$ with domain boundaries parallel to $\{10\bar{1}1\}$, also describes Brazil twins in low-quartz (α-SiO$_2$) (e.g., [25,26]). Similar twins involving reflection on planes perpendicular to the crystallographic **a** axis, like $\{11\bar{2}0\}$, have also been suggested for SBBO [27], which is structurally similar to KABO [13,14]. Note that for the space groups of YAB (R32 [1]), KABO (P321 [13]), and low-quartz (P3$_1$21), these twin laws generate merohedral twins [24] (twinning by merohedry [28]) that are not readily detected by conventional X-ray diffraction techniques. As these space groups share the same acentric and enantiomorphous crystal class (32), however, reflecting the crystal structure on planes perpendicular to the **a** axes not only inverts the polar two-fold axes along **a** but also converts the right-handed polytype of each structure into its left-handed counterpart and vice versa.

The domain structures of Yb:YAB and Nd:YAB have been shown to severely affect the SHG output at 532 nm by partially back-converting second harmonic light along interfaces between domains of opposing polarity [29,30]. Stacks of periodically inverted polar domains can, however, convert light to higher harmonics via quasi-phase-matching [31] as experimentally demonstrated using, for example, artificially twinned low-quartz crystals [32,33]. Indeed, accidental quasi-phase-matching has also been observed in Yb:YAB and Nd:YAB crystals with as-grown twin domains of suitable shape [29,30]. For crystals of class 32, inversion twins also affect the optical activity of twinned crystals as right- and left-handed domains rotate the electric field vector in opposing senses. To relate an observed optical rotation to inversion twinning, however, the unaffected or intrinsic rotatory power of the crystal species needs to be known. We determined the intrinsic rotatory powers of YAB and KABO as a function of the wavelength using a spectrophotometer with rotatable polarizers and applied the results to characterize the intracrystalline microstructure of these borates. Within individual YAB and KABO crystals, we found changes in the sign and magnitude of the optical rotation revealing inversion twins composed of right- and left-handed domains.

2. Materials and Methods

2.1. Crystal Growth and Sample Preparation

YAB crystals were grown by the top-seeded solution growth (TSSG) technique from YAB solutions based on the solvent systems Li$_2$O–WO$_3$–B$_2$O$_3$ [34] and Li$_2$O–Al$_2$O$_3$–B$_2$O$_3$ [8]. Depending on the solvent, the habitus of the YAB crystals varied between stubby and prismatic with the positive rhombohedron $\{10\bar{1}1\}$ and the trigonal prisms $\{11\bar{2}0\}$ and $\{\bar{1}1\bar{2}0\}$ as dominating crystal faces. In all YAB crystals, a central zone of inclusions extended along the pulling direction. A KABO crystal was synthesized by TSSG from a NaF solution [35,36] with a small excess of B$_2$O$_3$. The crystal did not develop flat crystal faces and contained polycrystalline regions similar to previous reports on KABO crystals grown from NaF flux [36]. Platelets with an orientation approximately perpendicular to the trigonal symmetry axis (crystallographic **c** axis) were cut from YAB and KABO crystals and ground down close to designated thicknesses using alumina powder. The platelets were then oriented perpendicular to the **c** axis to better than 9′ using X-ray diffraction in a Laue backscattering geometry and double-sided polished to plane-parallel sections using colloidal silica slurry. Section thicknesses were determined with a digital micrometer giving an accuracy of about 0.005 mm.

2.2. Measurement of the Rotatory Power at Different Wavelengths

In an optically active crystal, an incident linearly polarized light wave splits into two elliptically polarized waves with opposite senses of rotation, left (L) and right (R), that propagate through the crystal with different velocities according to their refractive indices n_L and n_R. At each point along the propagation path through the crystal, the different wavelengths of the two elliptically polarized waves give rise to a phase shift 2ϕ between the angles by which the dielectric displacement vectors of the two waves have been rotated. The phase shift 2ϕ is proportional to the difference in refractive indices and increases with the optical path length L through the crystal. As a result, the polarization of the combined wave that leaves the crystal is rotated by an angle ϕ with respect to the polarization of the incident light wave (e.g., [37,38]):

$$\phi = \frac{\pi(n_{\mathrm{L}} - n_{\mathrm{R}})L}{\lambda} \tag{1}$$

where λ is the wavelength of the light wave outside the crystal. The rotatory power or specific rotation is defined as [39]:

$$\rho = \phi/L = \frac{\pi(n_{\mathrm{L}} - n_{\mathrm{R}})}{\lambda} \tag{2}$$

For the definition of the sense of rotation and the sign of the rotatory power, see the standard reference by Glazer and Stadnicka [39]. YAB, KABO, and low-quartz are optically uniaxial, and we focus here on the optical activity for light propagation along the optical axis, i.e., parallel to the crystallographic **c** axis.

To measure the angle of rotation ϕ as a function of wavelength λ, we used a PerkinElmer Lambda 1050 spectrophotometer equipped with rotatable calcite polarizers. A schematic draft of the setup is shown in Figure 1a. Polychromatic light emitted by deuterium ($\lambda < 320$ nm) and tungsten ($\lambda > 320$ nm) lamps was sent through a double-stage monochromator based on holographic gratings and polarized by a motorized rotatable polarizer. A series of apertures was used to collimate the light beam to an effective aperture of about 1 mm on the crystal surface. The light was transmitted through the plane-parallel crystal section and through a second polarizer placed behind the crystal, the analyzer, and was detected with a photomultiplier ($\lambda < 820$ nm) or an InGaAs photodiode ($\lambda > 820$ nm). Polarizers were crossed by setting the rotation angle of the first polarizer to $\psi = 0$ and by manually rotating the analyzer so as to minimize the transmitted intensity. We crossed the polarizers for a wavelength of $\lambda = 1100$ nm to avoid bias by ambient light that might have entered the spectrophotometer during the alignment procedure.

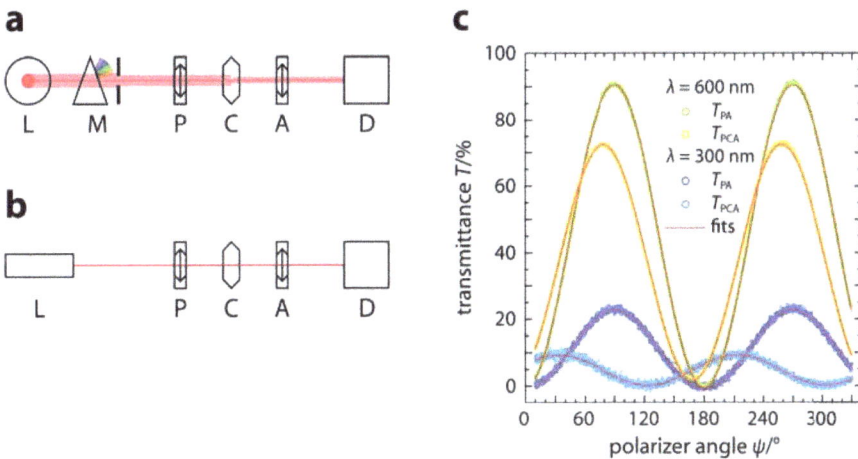

Figure 1. Experimental setup for the measurement of the optical rotatory power. (**a,b**) Arrangement of optical components in the spectrophotometer and for a setup using a laser as the light source. L: light source (deuterium or tungsten lamp in (**a**) and He-Ne laser in (**b**)), M: monochromator, P: rotatable polarizer, C: optically active crystal, A: rotatable analyzer, D: detector. (**c**) Transmittance through polarizer and analyzer (PA) and through polarizer, YAB crystal, and analyzer (PCA) recorded with the spectrophotometer as a function of the polarizer angle.

At a given wavelength, the polarizer angle ψ was scanned from $10°$ to $330°$ in steps of $0.15°$ and the light intensity T^*_{PCA} transmitted through polarizer (P), crystal (C), and analyzer (A) was recorded for 0.2 s at each step. Then, the crystal was removed from the beam path and the scan was repeated to measure the light intensity T^*_{PA} transmitted through polarizer and analyzer only. Transmitted light intensities T^* were recorded as transmittances relative to a reference beam path and were corrected for the inherent polarization of the incident beam and potential polarization-dependent sensitivities of the detectors using an additional polarization scan with only the polarizer in the beam path (T^*_P). Observed transmittances T^* were corrected as follows:

$$T_{\text{PCA}} = \frac{T^*_{\text{PCA}} - <T^*_0>}{T^*_\text{P} - <T^*_0>} \quad \text{and} \quad T_{\text{PA}} = \frac{T^*_{\text{PA}} - <T^*_0>}{T^*_\text{P} - <T^*_0>} \tag{3}$$

where $<T^*_0>$ is the mean transmittance for a polarization scan with the beam path blocked. Figure 1c shows the corrected transmittances T_{PCA} through a YAB crystal at 600 nm and 300 nm as a function of the polarizer angle ψ together with the corrected transmittance scans $T_{\text{PA}}(\psi)$ without the crystal section.

For each wavelength, the corrected transmittance scans $T_{\text{PCA}}(\psi)$ and $T_{\text{PA}}(\psi)$ were fit to functions of the form

$$T(\psi) = \Delta T \sin^2(\psi - \epsilon) + T(\epsilon) \tag{4}$$

with the polarizer extinction angle ϵ at which the transmittance is reduced to the minimum value $T(\epsilon)$ of the respective scan. ΔT measures the amplitude of the variation in transmittance along a polarization scan. Note that the three parameters ϵ, $T(\epsilon)$, and ΔT are constrained by more than 2000 data points of each transmittance scan. The angle ϕ by which the plane of polarization of the incident light beam is rotated by the optically active crystal is equal to the absolute difference $|\epsilon_{\text{PCA}} - \epsilon_{\text{PA}}|$. Accordingly, the magnitude of the rotatory power is given as

$$|\rho| = |\epsilon_{\text{PCA}} - \epsilon_{\text{PA}}|/L. \tag{5}$$

To determine the optical rotatory dispersion $\rho(\lambda)$, the measurements and analysis were repeated as described above for incident light of different wavelengths between 300 nm and 900 nm. Based on repeated measurements at the same wavelength, we estimate observed rotatory powers to be reproducible within $0.1°$ mm^{-1}.

Our approach of measuring the optical rotation angle as a function of wavelength shares some similarities with recently reported methods that are either based on channeled spectra recorded using a spectrophotometer with crossed polarizers [40,41] or that employ a motorized rotatable analyzer followed by a transmission grating to disperse the light onto a linear CCD array for simultaneous detection over a range of wavelengths [42]. Here, we combined the high spectral resolution and wide spectral range of the spectrophotometer with the high angular precision provided by polarization scans. By following the procedure outlined above, the transmission properties of a given pair of polarizers are readily calibrated without need to accurately cross them by hand.

2.3. Space-Resolved Measurements of the Optical Rotation

To detect potential variations in optical rotation within individual crystals, plane-parallel crystal sections were mounted onto a three-axis stage and aligned with their polished faces perpendicular to the beam of a He-Ne laser with 2 mW output power. Rotatable Glan–Thompson polarizers were placed in the beam path before (polarizer) and after (analyzer) the crystal section. Light transmitted through the polarizer, crystal, and analyzer was detected with a power meter. Figure 1b shows a scheme of the setup. The analyzer extinction angle ϵ was determined with and without the crystal section in the beam path by manually rotating the analyzer so as to minimize the transmitted light intensity by reading off the angle from the vernier scale on the analyzer mount. Again, the difference $|\epsilon_{\text{PCA}} - \epsilon_{\text{PA}}|$ equals the angle ϕ by which the crystal rotates the plane of polarization of the laser. By moving the crystal with the three-axis stage, the rotation angle ϕ was measured at different positions through the crystal section. The diameter of the laser beam limited the lateral spatial resolution to about 1 mm. Repeated measurements that aimed to reproduce both the position on the crystal section as well as the extinction angle ϵ at that position showed the observed rotation angles ϕ to be precise within $0.1°$.

3. Results

3.1. Low-Quartz, α-SiO$_2$

An oriented low-quartz crystal with plane-parallel polished faces perpendicular to the crystallographic **c** axis and a thickness of 4.817(5) mm served as a reference to assess the accuracy of our measurements of the rotatory power at different wavelengths using the spectrophotometer. Figure 2 compares our results with previous measurements [43] and calculations [44] of the rotatory power of low-quartz. We described the optical rotatory dispersion using a semi-empirical formalism based on the theory of circular dichroism [39,45]:

$$\rho = A/(\lambda^2 - \lambda_1^2). \tag{6}$$

The constant A captures the strength of dispersion, while λ_1 is the wavelength of an optically active electronic transition [39]. While more elaborate formulations for the optical rotatory dispersion of low-quartz have been proposed [45–47], we found the simple expression (6) to accurately describe our experimental data within their uncertainties. By fitting Equation (6) to our experimental results, we obtained $A = 7.17(3)°$ mm and $\lambda_1 = 129(1)$ nm. Inter- and extrapolating our results using these parameters in the dispersion relation (6) gave excellent agreement with literature data on the rotatory power of low-quartz, even at wavelengths in the deep UV where the rotatory power is strongly affected by dispersion and outside the spectral range accessible with the spectrophotometer (Figure 2). This result demonstrates the accuracy of our method to determine the rotatory power at different wavelengths using a spectrophotometer.

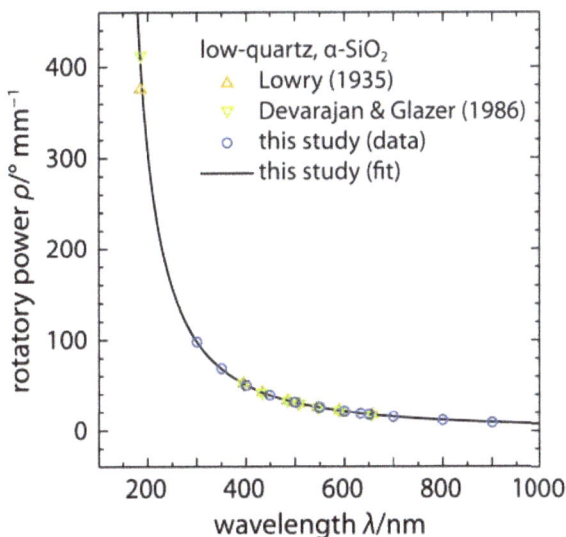

Figure 2. Optical rotatory dispersion of low-quartz, α-SiO$_2$. Lowry (1935): measurements [43]; Devarajan & Glazer (1986): calculation [44]; this study: measurements and fit to data.

3.2. YAB, YAl₃(BO₃)₄

In the case of YAB, the measurement of the rotatory power is complicated by inversion twins that might reduce the observed rotatory power to values below the intrinsic rotatory power. When a light wave traverses a sequence of left- and right-handed domains, their individual contributions to the observed optical rotation angle ϕ will partly cancel each other out and thereby result in underestimation of the rotatory power. We searched for a YAB crystal containing a large tentative single domain that provided the longest possible optical path length along the **c** axis. We then measured the optical rotation angle through this domain at different wavelengths. The crystal was then reground and polished to shorten the optical path length along the **c** axis without changing the crystal orientation, and the measurement of the optical rotation angle was repeated. In total, we measured optical rotation angles at 11 different wavelengths for five different optical path lengths on the same domain. In case the domain contained any twin lamellae with opposite senses of rotation, every change in the optical path length should have changed the relative proportions of left- and right-handed domains along the optical path and hence, the observed rotatory power. For the selected domain, we did not observe significant changes of the rotatory power when the optical path length changed.

Figure 3a shows the measured optical rotation angles ϕ for different wavelengths as a function of the optical path length L. For a given wavelength, the observed rotation angles display a linear relation with the optical path length. In particular, the highest rotation angles are observed for the longest optical path lengths. These observations suggest that the probed crystal volume was composed of a single domain and that the observed rotation angles were not affected by lamellae of inversion twins. To extract the dispersion relation of the rotatory power of YAB, we analyzed the complete data set of pairs of rotation angles ϕ and optical path lengths L using a modified version of Equation (6):

$$\phi = AL/(\lambda^2 - \lambda_1^2). \tag{7}$$

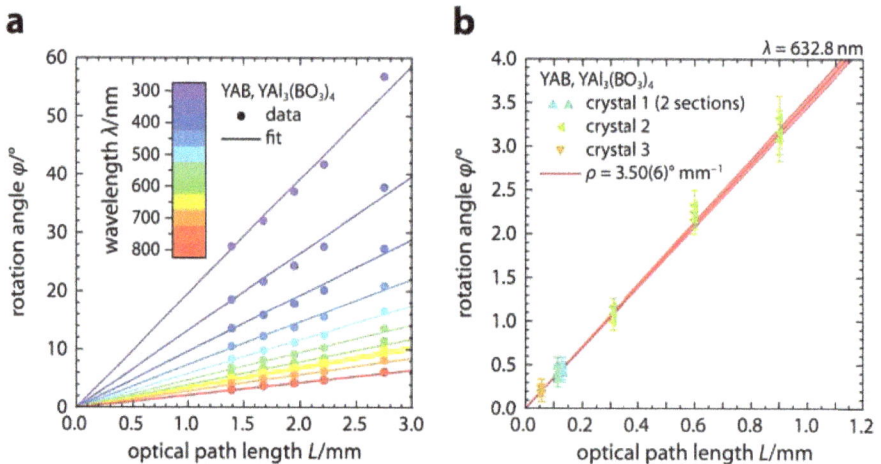

Figure 3. Optical rotation angles of YAB measured on domains with different optical path lengths. In (**a**), the lines show the result of simultaneously fitting a dispersion relation to the observed rotation angles at all wavelengths and path lengths shown in (**a**). The red line in (**b**) shows the fitting result of (**a**) for a wavelength of 632.8 nm. The shading indicates propagated uncertainties on fitted parameters. Note the linear relation between observed rotation angles and path lengths in (**a**) and the agreement of the fit with measurements at 632.8 nm on very thin domains in (**b**).

A fit to all data in Figure 3a gave $A = 1.32(2)°$ mm and $\lambda_1 = 149(4)$ nm. The corresponding dispersion curve is shown in Figure 4 together with rotatory powers, as calculated from the observed rotation angles.

To consolidate the results on the rotatory power of YAB obtained from measurements on a single large domain using the spectrophotometer, we searched for single domains of suitable size to be probed using the experimental setup with a He-Ne laser as the light source (see Section 2.3 and Figure 1b). The combination of the spatially confined laser beam with the three-axis stage allowed us to precisely navigate to and determine the optical rotation angles on small crystal volumes. Figure 3b compiles the results of measurements on single domains distributed over four crystal sections that were cut from three different YAB crystals. Note the good agreement between measurements on small domains with short optical path lengths using the laser setup and those on the large domain performed with the spectrophotometer. At the wavelength of the He-Ne laser ($\lambda = 632.8$ nm), the parametrization of the dispersion relation of YAB derived above gives an optical rotatory power of $\rho(632.8 \text{ nm}) = 3.50(6)°$ mm^{-1}.

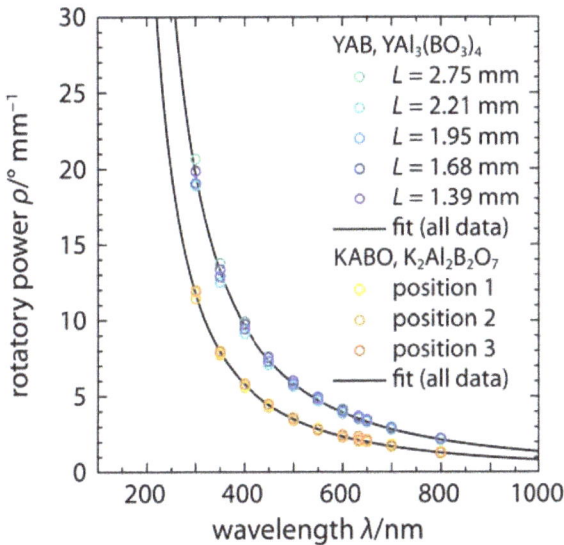

Figure 4. Optical rotatory dispersion of YAB (YAl$_3$(BO$_3$)$_4$) and KABO (K$_2$Al$_2$B$_2$O$_7$). For YAB, rotatory powers were measured on the same domain for five different optical path lengths (see also Figure 3a). For KABO, rotatory powers were measured at three different locations on the same crystal section.

3.3. KABO, $K_2Al_2B_2O_7$

The rotatory power of a KABO crystal with a thickness of 0.717(5) mm was measured along the **c** axis at 11 different wavelengths and at three different locations. The results are shown in Figure 4 together with the rotatory powers of YAB. Analysis of the KABO data with Equation (6) gave $A = 0.80(1)°$ mm and $\lambda_1 = 148(3)$ nm. Based on this dispersion relation, KABO rotates the plane of polarization by $2.10(3)°$ mm^{-1} at a wavelength of $\lambda = 632.8$ nm. The parameters for the dispersion relations of low-quartz, YAB, and KABO are compiled in Table 1.

Table 1. Dispersion relations for the optical rotatory power of low-quartz, YAB, and KABO.

Crystal	Dispersion Relation [a]		
	A (° mm)	λ_1 (nm)	ρ_{633}^b (° mm^{-1})
Low-quartz, α-SiO$_2$	7.17(3)	129(1)	18.69(9)
YAB, YAl$_3$(BO$_3$)$_4$	1.32(2)	149(4)	3.50(6)
KABO, K$_2$Al$_2$B$_2$O$_7$	0.80(1)	148(3)	2.10(3)

[a] $\rho = A/(\lambda^2 - \lambda_1^2)$; [b] rotatory power at $\lambda = 632.8$ nm.

4. Discussion

4.1. Relation between Optical Activity and Crystal Structure

From a microscopic point of view, optical activity arises from a chiral arrangement of polarizable electron density. In crystals of the enantiomorphous crystal classes, the periodicity of the atomic structure often results in helical arrangements of atoms or larger structural units such as coordination polyhedra. The geometrical characteristics of such helices have been related to the sign and magnitude of the rotatory power in a number of inorganic compounds [39,44,48]. The dispersion of the rotatory power arises from transitions between electronic states that involve those orbitals that host the optically active electron density [39]. The wavelength or energy of optically active transitions therefore contains information on the structural units that contribute to the optical activity.

The crystal structure of low-quartz is composed of [SiO$_4$]$^{4-}$ tetrahedra that share corners to form helices along the **c** axis. Indeed, the absorption wavelength derived from the dispersion relation of the rotatory power of low-quartz (Table 1) agrees well with observed and computed energy separations of about 10 eV between occupied and unoccupied orbitals of [SiO$_4$]$^{4-}$ tetrahedra [49]. The crystal structures of YAB and KABO are more complex. In YAB, helices of [AlO$_6$]$^{9-}$ octahedra wind along the **c** axis and are linked to each other by planar [BO$_3$]$^{3-}$ groups and slightly twisted [YO$_6$]$^{9-}$ prisms [50]. In addition to planar [BO$_3$]$^{3-}$ groups, the polyhedral framework of KABO contains tetrahedral [AlO$_4$]$^{5-}$ groups and irregularly coordinated K$^+$ cations [13]. The absorption wavelengths in the dispersion relations of both YAB and KABO compare best with predicted energy gaps between occupied and unoccupied orbitals of the [BO$_3$]$^{3-}$ group with a magnitude slightly above 8 eV [4,51]. Electronic transitions in octahedral [AlO$_6$]$^{9-}$ and tetrahedral [AlO$_4$]$^{5-}$ groups are expected at higher energies [49,52]. We therefore conclude that the optical activity of YAB and KABO arises mainly from the arrangement of [BO$_3$]$^{3-}$ groups. Since most of the polarizable electron density concentrates around oxygen atoms that are shared with other structural units, a more complete approach to explain the optical activity should take into account the crystal structure as a whole [39,44].

We further note that the dispersion of the rotatory power might provide an alternative way to estimate the intrinsic absorption edge of optically active crystals. Even small amounts of impurities or defects may cause strong absorption of light at energies below the absorption edge and may therefore mask the absorption edge in conventional spectroscopic experiments. In contrast, optically active electronic transitions follow different selection rules [39] that limit the influence of impurities on the dispersion of the rotatory power. In YAB and KABO, for example, substantial absorption of UV radiation has been observed at energies well below the expected absorption edge and is attributed to impurities of Fe^{3+} [7,8,53]. As a result, YAB and KABO crystals become effectively nontransparent at wavelengths below 170 nm and 180 nm, respectively [8,53]. Based on the dispersion of the rotatory power, however, we tentatively locate the intrinsic absorption edge at around 150 nm (Table 1), in excellent agreement with first-principle calculations on the electronic structure of the [BO$_3$]$^{3-}$ group [4,51].

4.2. Changes in Sign and Magnitude of Optical Activity within YAB and KABO Crystals

Figure 5 shows sections of YAB and KABO crystals cut perpendicular to the **c** axis, i.e., perpendicular to the the optical axis. The sections were viewed between crossed polarizers. To generate a visible contrast between right- and left-handed domains, however, the analyzer was rotated to bring one type of domain into extinction, a technique also used to visualize Brazil twins in low-quartz [25,26]. In YAB (Figure 5a), domain boundaries are inclined relative to the **c** axis and appear blurred. In the KABO section, domain boundaries give rise to sharp contrasts (Figure 5b), indicating an orientation of the boundaries parallel to the **c** axis.

To demonstrate that the contrast between domains in the images indeed arises from opposing signs of optical activity, we measured the apparent rotatory power ρ^* at different locations on the crystal sections using the setup with the He-Ne laser (Section 2.3 and Figure 1b). The results are superimposed on the images as the color-coded ratio of apparent to intrinsic rotatory power $\rho^*/|\rho|$ at a wavelength of 632.8 nm. A spatial correlation exists both between domains in extinction (dark) and positive apparent rotatory powers (orange) and between bright regions and negative apparent rotatory powers (blue). Changes in sign of the apparent rotatory power across the crystal sections demonstrate the presence of right- and left-handed domains within individual YAB and KABO crystals. As proposed for YAB [18,21], we interpret these domains as inversion twins and note, to the best of our knowledge, that inversion twins have not previously been detected in KABO crystals.

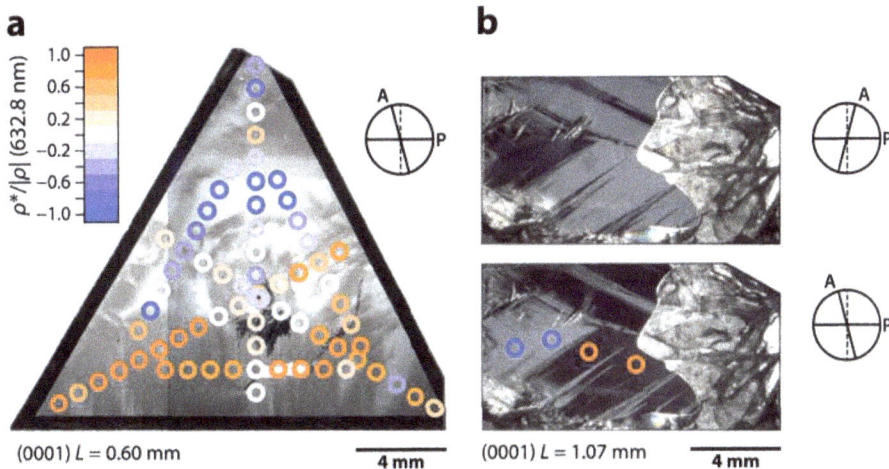

Figure 5. (**a**) YAB and (**b**) KABO crystals cut perpendicular to the **c** axis. The crystal sections are viewed between slightly decrossed polarizers as schematically indicated by the diagrams showing the relative orientations of the polarizer (P) and analyzer (A) for each image. Note the congruence of domains that are in extinction positions in the upper panel of (**b**) and appear bright in the lower panel and vice versa. The colored rings give the positions for measurements of the apparent rotatory power. The size of the symbols reflects the approximate diameter of the laser beam. Rings are colored according to the relative apparent rotatory power at a wavelength of 632.8 nm, as indicated by the color scale in (**a**). Note the correlation between the extinction behavior and the sign of the apparent rotatory power.

While dark and bright domains in the KABO section give apparent rotatory powers close to 1 and −1, respectively, a whole range of intermediate values is observed in the YAB section. Reduced magnitudes of the apparent rotatory power are best explained by contributions of both right- and left-handed domains to the crystal volume probed by the laser beam. Fine lamellar intergrowths of the right- and left-handed domains can be found in both crystal sections (Figure 5). In the case of YAB, this microstructure most certainly corresponds to polysynthetic inversion twinning, which was reported in earlier studies using different imaging techniques [18–21,24]. Domain boundaries that are inclined with respect to the light path may also lead to a superposition of wedges with opposing signs of rotation and thereby reduce the observed rotation angle. In principle, the magnitude of the ratio $\rho^*/|\rho|$ corresponds to the volume fraction of one type of domain within the probed crystal volume. Any quantitative estimate would require knowing the exact geometry and internal intensity distribution of the light beam inside the crystal, a task we did not attempt here. Our descriptions for the dispersion relations of the rotatory power of YAB and KABO, however, facilitate such estimates to be done at different wavelengths, offering an inexpensive and practicable method to assess the extent of inversion twinning.

Depending on the actual twin law, a pattern of domains with exactly opposing polarities can be formed in different ways. In crystal class 32, a reflection on planes perpendicular to the polar **a** axes will result in a pattern with domains of exactly opposing polarities and, at the same time, convert the right-handed polytype into the left-handed one (and vice versa). This would correspond to so-called Brazil twins in low-quartz. As with Dauphiné twins in low-quartz, domains with opposing polarities could also result from twinning by a two-fold rotation around the **c** axis. Such a two-fold rotation around the **c** axis, however, would not invert the sense of optical rotation. The domains visible in the YAB and KABO crystals shown in Figure 5 are therefore inversion twins that require reflection as twin operation. Our observations of right- and left-handed domains complement earlier analyses of the microstructure of YAB crystals that proposed reflection on $\{11\bar{2}0\}$ planes as twin operation [18,21]. This twin law generates domains with opposing polarities, and the resulting microstructure is expected to strongly affect the NLO properties of twinned YAB crystals [21,24,30].

5. Conclusions

We developed and tested a simple method to measure the rotatory power of optically active crystals at different wavelengths using a spectrophotometer with rotatable polarizers. This method was applied to determine the optical rotatory dispersion of the NLO borate crystals YAB and KABO. In both YAB and KABO, the absorption wavelengths of optically active electronic absorptions reflect the electronic structure of the $[BO_3]^{3-}$ group, suggesting that the optical activity arises mainly from the arrangement of borate groups in the crystal structures of these materials. By probing different locations in oriented crystal sections with a He-Ne laser beam, we found the sign and magnitude of the optical rotation to vary within individual YAB and KABO crystals. The variations in apparent rotatory power correlate with the microstructure of the crystals when viewed between crossed polarizers. We explain these changes in sign and magnitude of the apparent rotatory power with right- and left-handed domains that form inversion twins. In combination with proposed twin laws, such inversion twins might affect the NLO properties of YAB and KABO crystals if polar axes of adjacent domains point in opposite directions. The here-derived dispersion relations for the rotatory power of YAB and KABO facilitate fast and inexpensive assessment of the extent of inversion twinning. Based on quantitative measurements of the rotatory power, a direct measurement of the degree of twinning of the investigated crystal volume and the selection of high-quality (twin-free) crystal specimens is now possible.

Author Contributions: J.B. and V.W. designed and developed the measurement procedures; J.B. and S.D. performed measurements and analyzed data; V.W., A.G., and D.R. grew YAB and KABO crystals; V.W. and D.R. supervised and managed the project; J.B. wrote the manuscript; all authors commented on the manuscript.

Funding: This research was funded in part by the Bundesministerium für Bildung und Forschung, grant number 13N11560.

Acknowledgments: We thank B. Wenzel, M. Meiers, and A. Klintz for sample preparation and optical polishing and S. Ilas, J. Ren, J. Lejay, P. Loiseau, G. Aka, R. Maillard, and A. Maillard for enlightening discussions. K. Dupré provided a low-quartz crystal, and M. Peltz built the He-Ne laser setup for measurements of the optical activity.

Conflicts of Interest: The authors declare no conflict of interest. The funders had no role in the design of the study; in the collection, analyses, or interpretation of data; in the writing of the manuscript, or in the decision to publish the results.

Abbreviations

The following abbreviations are used in this manuscript:

BBO	β-BaB$_2$O$_4$
CLBO	CsLiB$_6$O$_{10}$
KABO	K$_2$Al$_2$B$_2$O$_7$
KBBF	KBe$_2$BO$_3$F$_2$
LBO	LiB$_3$O$_5$
NLO	nonlinear optical
REE	rare-earth element
SBBO	Sr$_2$Be$_2$B$_2$O$_7$
SHG	second harmonic generation
TSSG	top-seeded solution growth
UV	ultraviolet
YAB	YAl$_3$(BO$_3$)$_4$
YLSB	Y$_{0.57}$La$_{0.72}$Sc$_{2.71}$(BO$_3$)$_4$

References

1. Mills, A.D. Crystallographic data for new rare earth borate compounds, RX$_3$(BO$_3$)$_4$. *Inorg. Chem.* **1962**, *1*, 960–961. [CrossRef]
2. Ballman, A.A. A new series of synthetic borates isostructural with the carbonate mineral huntite. *Am. Mineral.* **1962**, *47*, 1380–1383.
3. Leonyuk, N.I.; Leonyuk, L.I. Growth and characterization of RM$_3$(BO$_3$)$_4$ crystals. *Prog. Cryst. Growth Charact.* **1995**, *31*, 179–278. [CrossRef]
4. Chen, C.; Wang, Y.; Xia, Y.; Wu, B.; Tang, D.; Wu, K.; Wenrong, Z.; Yu, L.; Mei, L. New development of nonlinear optical crystals for the ultraviolet region with molecular engineering approach. *J. Appl. Phys.* **1995**, *77*, 2268–2272. [CrossRef]
5. Becker, P. Borate materials in nonlinear optics. *Adv. Mater.* **1998**, *10*, 979–992. [CrossRef]
6. Xue, D.; Betzler, K.; Hesse, H.; Lammers, D. Nonlinear optical properties of borate crystals. *Solid State Commun.* **2000**, *114*, 21–25. [CrossRef]
7. Yu, X.; Yue, Y.; Yao, J.; Hu, Z.G. YAl$_3$(BO$_3$)$_4$: Crystal growth and characterization. *J. Cryst. Growth* **2010**, *312*, 3029–3033. [CrossRef]
8. Yu, J.; Liu, L.; Zhai, N.; Zhang, X.; Wang, G.; Wang, X.; Chen, C. Crystal growth and optical properties of YAl$_3$(BO$_3$)$_4$ for UV applications. *J. Cryst. Growth* **2012**, *341*, 61–65. [CrossRef]
9. Rytz, D.; Gross, A.; Vernay, S.; Wesemann, V. YAl$_3$(BO$_3$)$_4$: A novel NLO crystal for frequency conversion to UV wavelengths. In Proceedings of the Solid State Lasers and Amplifiers III SPIE, Strasbourg, France, 7–11 April 2008; Volume 6998, p. 699814. [CrossRef]
10. Chen, C.T.; Wu, B.C.; Jiang, A.D.; You, G.M. A new-type ultraviolet SHG crystal: β-BaB$_2$O$_4$. *Sci. Sin. B* **1985**, *18*, 235–243.
11. Chen, C.; Wu, Y.; Jiang, A.; Wu, B.; You, G.; Li, R.; Lin, S. New nonlinear-optical crystal: LiB$_3$O$_5$. *J. Opt. Soc. Am. B* **1989**, *6*, 616–621. [CrossRef]
12. Mori, Y.; Kuroda, I.; Nakajima, S.; Sasaki, T.; Nakai, S. New nonlinear optical crystal: Cesium lithium borate. *Appl. Phys. Lett.* **1995**, *67*, 1818–1820. [CrossRef]
13. Hu, Z.G.; Higashiyama, T.; Yoshimura, M.; Yap, Y.K.; Mori, Y.; Sasaki, T. A new nonlinear optical borate crystal K$_2$Al$_2$B$_2$O$_7$ (KAB). *Jpn. J. Appl. Phys.* **1998**, *37*, L1093. [CrossRef]
14. Ye, N.; Zeng, W.; Jiang, J.; Wu, B.; Chen, C.; Feng, B.; Zhang, X. New nonlinear optical crystal K$_2$Al$_2$B$_2$O$_7$. *J. Opt. Soc. Am. B* **2000**, *17*, 764. [CrossRef]
15. Chen, C.; Wang, Y.; Wu, B.; Wu, K.; Zeng, W.; Yu, L. Design and synthesis of an ultraviolet-transparent nonlinear optical crystal Sr$_2$Be$_2$B$_2$O$_7$. *Nature* **1995**, *373*, 322–324. [CrossRef]
16. Chen, C.; Xu, Z.; Deng, D.; Zhang, J.; Wong, G.K.L.; Wu, B.; Ye, N.; Tang, D. The vacuum ultraviolet phase-matching characteristics of nonlinear optical KBe$_2$BO$_3$F$_2$ crystal. *Appl. Phys. Lett.* **1996**, *68*, 2930–2932. [CrossRef]

17. Wu, B.; Tang, D.; Ye, N.; Chen, C. Linear and nonlinear optical properties of the $KBe_2BO_3F_2$ (KBBF) crystal. *Opt. Mater.* **1996**, *5*, 105–109. [CrossRef]

18. Hu, X.B.; Jiang, S.S.; Huang, X.R.; Liu, W.J.; Ge, C.Z.; Wang, J.Y.; Pan, H.F.; Ferrari, C.; Gennari, S. X-ray topographic study of twins in $Nd_xY_{(1-x)}Al_3(BO_3)_4$ crystal. *Nuovo Cimento D* **1997**, *19*, 175–180. [CrossRef]

19. Hu, X.B.; Jiang, S.S.; Huang, X.R.; Liu, W.J.; Ge, C.Z.; Wang, J.Y.; Pan, H.F.; Jiang, J.H.; Wang, Z.G. The growth defects in self-frequency-doubling laser crystal $Nd_xY_{1-x}Al_3(BO_3)_4$. *J. Cryst. Growth* **1997**, *173*, 460–466. [CrossRef]

20. Péter, A.; Polgár, K.; Beregi, E. Revealing growth defects in non-linear borate single crystals by chemical etching. *J. Cryst. Growth* **2000**, *209*, 102–109. [CrossRef]

21. Zhao, S.; Wang, J.; Sun, D.; Hu, X.; Liu, H. Twin structure in $Yb:YAl_3(BO_3)_4$ crystal. *J. Appl. Crystallogr.* **2001**, *34*, 661–662. [CrossRef]

22. Ye, N.; Stone-Sundberg, J.L.; Hruschka, M.A.; Aka, G.; Kong, W.; Keszler, D.A. Nonlinear optical crystal $Y_xLa_ySc_z(BO_3)_4$ $(x + y + z = 4)$. *Chem. Mater.* **2005**, *17*, 2687–2692. [CrossRef]

23. Bourezzou, M.; Maillard, A.; Maillard, R.; Villeval, P.; Aka, G.; Lejay, J.; Loiseau, P.; Rytz, D. Crystal defects revealed by Schlieren photography and chemical etching in nonlinear single crystal LYSB. *Opt. Mater. Express* **2011**, *1*, 1569–1576. [CrossRef]

24. Ilas, S. Elaboration et Caractérisation de Matériaux Non-Linéaires Pour la Conception de Dispositifs Laser Émettant Dans L'ultraviolet. Ph.D. Thesis, Université Pierre et Marie Curie-Paris VI, Paris, France, 2014.

25. Schlössin, H.H.; Lang, A.R. A study of repeated twinning, lattice imperfections and impurity distribution in amethyst. *Philos. Mag.* **1965**, *12*, 283–296. [CrossRef]

26. McLaren, A.C.; Pitkethly, D.R. The twinning microstructure and growth of amethyst quartz. *Phys. Chem. Miner.* **1982**, *8*, 128–135. [CrossRef]

27. He, M.; Kienle, L.; Simon, A.; Chen, X.L.; Duppel, V. Re-examination of the crystal structure of $Na_2Al_2B_2O_7$: stacking faults and twinning. *J. Solid State Chem.* **2004**, *177*, 3212–3218. [CrossRef]

28. Giacovazzo, C.; Monaco, H.L.; Artioli, G.; Viterbo, D.; Milanesio, M.; Gilli, G.; Gilli, P.; Zanotti, G.; Ferraris, G.; Catti, M. (Eds.) *Fundamentals of Crystallography*, 3rd ed.; Oxford University Press: Oxford, UK, 2011.

29. Dekker, P.; Dawes, J.M. Characterisation of nonlinear conversion and crystal quality in Nd- and Yb-doped YAB. *Opt. Express* **2004**, *12*, 5922–5930. [CrossRef]

30. Dekker, P.; Dawes, J. Twinning and "natural quasi-phase matching" in Yb:YAB. *Appl. Phys. B* **2006**, *83*, 267. [CrossRef]

31. Fejer, M.M.; Magel, G.A.; Jundt, D.H.; Byer, R.L. Quasi-phase-matched second harmonic generation: tuning and tolerances. *IEEE J. Quantum Elect.* **1992**, *28*, 2631–2654. [CrossRef]

32. Kurimura, S.; Harada, M.; Muramatsu, K.i.; Ueda, M.; Adachi, M.; Yamada, T.; Ueno, T. Quartz revisits nonlinear optics: Twinned crystal for quasi-phase matching [Invited]. *Opt. Mater. Express* **2011**, *1*, 1367. [CrossRef]

33. Ishizuki, H.; Taira, T. Quasi phase-matched quartz for intense-laser pumped wavelength conversion. *Opt. Express* **2017**, *25*, 2369. [CrossRef]

34. Liu, H.; Li, J.; Fang, S.H.; Wang, J.Y.; Ye, N. Growth of $YAl_3(BO_3)_4$ crystals with tungstate based flux. *Mater. Res. Innov.* **2011**, *15*, 102–106. [CrossRef]

35. Zhang, C.; Wang, J.; Hu, X.; Liu, H.; Wei, J.; Liu, Y.; Wu, Y.; Chen, C. Top-seeded growth of $K_2Al_2B_2O_7$. *J. Cryst. Growth* **2001**, *231*, 439–441. [CrossRef]

36. Zhang, C.; Wang, J.; Hu, X.; Jiang, H.; Liu, Y.; Chen, C. Growth of large $K_2Al_2B_2O_7$ crystals. *J. Cryst. Growth* **2002**, *235*, 1–4. [CrossRef]

37. Nye, J.F. *Physical Properties of Crystals: Their Representation by Tensors and Matrices*; Oxford University Press: Oxford, UK, 1985; p. 352.

38. Haussühl, S. *Physical Properties of Crystals: An Introduction*; Wiley-VCH: Weinheim, Germany, 2007; p. 453.

39. Glazer, A.M.; Stadnicka, K. On the origin of optical activity in crystal structures. *J. Appl. Crystallogr.* **1986**, *19*, 108–122. [CrossRef]

40. Dimitriu, D.G.; Dorohoi, D.O. New method to determine the optical rotatory dispersion of inorganic crystals applied to some samples of Carpathian Quartz. *Spectrochim. Acta Part A* **2014**, *131*, 674–677. [CrossRef] [PubMed]

41. Dorohoi, D.O.; Dimitriu, D.G.; Cosutchi, I.; Breaban, I.; Closca, V. A new method for determining the optical rotatory dispersion of transparent crystalline layers. In Proceedings of the Second International Conference on Applications of Optics and Photonics, Aveiro, Portugal, 26–30 May 2014; Volume 9286, p. 92862Z. [CrossRef]

42. Jiang, S.; Jia, H.; Lei, Y.; Shen, X.; Cao, J.; Wang, N. Novel method for determination of optical rotatory dispersion spectrum by using line scan CCD. *Opt. Express* **2017**, *25*, 7445–7454. [CrossRef] [PubMed]

43. Lowry, T.M. *Optical Rotatory Power*; Longmans: London, UK, 1935.

44. Devarajan, V.; Glazer, A.M. Theory and computation of optical rotatory power in inorganic crystals. *Acta Crystallogr. A* **1986**, *42*, 560–569. [CrossRef]

45. Chandrasekhar, S. Optical rotatory dispersion of crystals. *Proc. R. Soc. Lond. Ser. A* **1961**, *259*, 531–553. [CrossRef]

46. Lowry, T.M.; Coode-Adams, W.R.C.X. Optical rotatory dispersion. Part III.—The rotatory dispersion of quartz in the infra-red, visible and ultra-violet regions of the spectrum. *Philos. Trans. R. Soc. A* **1927**, *226*, 391–466. [CrossRef]

47. Katzin, L.I. The rotatory dispersion of quartz. *J. Phys. Chem.* **1964**, *68*, 2367–2370. [CrossRef]

48. Ramachandran, G.N. Theory of optical activity of crystals. In *Proceedings of the Indian Academy of Sciences-Section A*; Indian Academy of Sciences: Bengaluru, India, 1951; Volume 33, pp. 217–227. [CrossRef]

49. Tossell, J.A. Electronic structures of silicon, aluminum, and magnesium in tetrahedral coordination with oxygen from SCF-Xα MO calculations. *J. Am. Chem. Soc.* **1975**, *97*, 4840–4844. [CrossRef]

50. Belokoneva, E.L.; Azizov, A.V.; Leonyuk, N.I.; Simonov, M.A.; Belov, N.V. Crystal structure of $YAl_3[BO_3]_4$. *J. Struct. Chem.* **1981**, *22*, 476–478. [CrossRef]

51. Tossell, J.A. Studies of unoccupied molecular orbitals of the B–O bond by molecular orbital calculations, X-ray absorption near edge, electron transmission, and NMR spectroscopy. *Am. Mineral.* **1986**, *71*, 1170–1177.

52. Tossell, J.A. The electronic structures of Mg, Al and Si in octahedral coordination with oxygen from SCF Xα MO calculations. *J. Phys. Chem. Solids* **1975**, *36*, 1273–1280. [CrossRef]

53. Liu, L.; Liu, C.; Wang, X.; Hu, Z.G.; Li, R.K.; Chen, C.T. Impact of Fe^{3+} on UV absorption of $K_2Al_2B_2O_7$ crystals. *Solid State Sci.* **2009**, *11*, 841–844. [CrossRef]

crystals

MDPI

Review

Crystallochemical Design of Huntite-Family Compounds

Galina M. Kuz'micheva [1], **Irina A. Kaurova** [1,*], **Victor B. Rybakov** [2] **and Vadim V. Podbel'skiy** [3]

[1]   MIREA-Russian Technological University, Vernadskogo pr. 78, Moscow 119454, Russia; kaurchik@yandex.ru
[2]   Lomonosov Moscow State University named M.V. Lomonosov, Vorobyovy Gory, Moscow 119992, Russia;
       rybakov20021@yandex.ru
[3]   National Research University «Higher School of Economics», Myasnitskaya str. 20,
       Moscow 101000, Russia; vpodbelskiy@hse.ru
*   Correspondence: kaurchik@yandex.ru; Tel.: +7-495-246-0555 (ext. 434)

Received: 18 December 2018; Accepted: 12 February 2019; Published: 15 February 2019

Abstract: Huntite-family nominally-pure and activated/co-activated $LnM_3(BO_3)_4$ (Ln = La–Lu, Y; M = Al, Fe, Cr, Ga, Sc) compounds and their-based solid solutions are promising materials for lasers, nonlinear optics, spintronics, and photonics, which are characterized by multifunctional properties depending on a composition and crystal structure. The purpose of the work is to establish stability regions for the rare-earth orthoborates in crystallochemical coordinates (sizes of Ln and M ions) based on their real compositions and space symmetry depending on thermodynamic, kinetic, and crystallochemical factors. The use of diffraction structural techniques to study single crystals with a detailed analysis of diffraction patterns, refinement of crystallographic site occupancies (real composition), and determination of structure–composition correlations is the most efficient and effective option to achieve the purpose. This approach is applied and shown primarily for the rare-earth scandium borates having interesting structural features compared with the other orthoborates. Visualization of structures allowed to establish features of formation of phases with different compositions, to classify and systematize huntite-family compounds using crystallochemical concepts (structure and superstructure, ordering and disordering, isostructural and isotype compounds) and phenomena (isomorphism, morphotropism, polymorphism, polytypism). Particular attention is paid to methods and conditions for crystal growth, affecting a crystal real composition and symmetry. A critical analysis of literature data made it possible to formulate unsolved problems in materials science of rare-earth orthoborates, mainly scandium borates, which are distinguished by an ability to form internal and substitutional (Ln and Sc atoms), unlimited and limited solid solutions depending on the geometric factor.

Keywords: huntite family; rare-earth scandium borate; optical material; crystal growth; rare-earth cations; X-ray diffraction (XRD); crystal structure; solid solution; order–disorder

---

1. Introduction

Modern scientific and applied materials science requires both an appearance of new materials with a desired combination of functional properties and an improvement and optimization of physical parameters and structural quality of the known materials, which have already proven themselves in practice, with further control of their properties using external (growth and post-growth treatment conditions) or internal (activation, isomorphic substitution) effects. Isomorphic substitutions and activation are different in the concentration of ion(s) introduced into a crystal matrix and effects produced. Isomorphic substitution is a powerful and flexible way to obtain a desired physical parameter of material by a targeted change in the composition of specific crystal structure. Activation is one of the necessary and relatively simple technological actions for the appearance, modification, and improvement of crystal properties by introduction (sometimes, over stoichiometry) of small amounts of impurity ions into a crystal matrix that contributes to a self-organization and self-compensation of system's electroneutrality. It is possible that dopant ions introduced into a crystal structure in low concentrations over stoichiometry are distributed over crystallographic sites in a different way than those introduced in high concentrations in the case of formation of solid solutions. However, low content of dopant ions can lead to significant changes in both local and statistical structures, in particular, in crystal symmetry. In this paper, all the above-mentioned aspects are reviewed for the huntite family compounds, interesting and important materials from both applied and scientific points of view, with the main focus on single-crystal objects, a detailed study of which can obtainin reliable information about the internal structure of materials.

Complex orthoborates of rare-earth metals with the general chemical formula $LnM_3(BO_3)_4$, where Ln^{3+} = La–Lu, Y and M^{3+} = Al, Ga, Sc, Cr, Fe, belong to the huntite family (huntite $CaMg_3(CO_3)_4$, space group $R32$ [1]). Depending on the composition and external conditions, they can have both monoclinic (space groups $C2/c$, Cc, and $C2$) and trigonal (space groups $R32$, $P321$, and $P3_12$) symmetry with the presence (space group $C2/c$) or absence of center of symmetry.

The most widely known and studied representatives of the huntite family compounds are aluminum borates $LnM_3(BO_3)_4$ with the M^{3+} = Al. The $LnAl_3(BO_3)_4$ crystals with the Ln^{3+} = Y, Gd, Lu doped with the Nd, Dy, Er, Yb, Tm (for example, [2,3]) and co-doped with the Er/Tb, Er/Yb, Nd/Yb ions (for example, [4,5]), as well as single-crystal solid solutions in the $YAl_3(BO_3)_4 – NdAl_3(BO_3)_4$ system [6], are promising materials for self-frequency-doubled lasers. $YAl_3(BO_3)_4$ crystals doped with the Er^{3+} or Yb^{3+} ions are widely used in medicine and telecommunications as laser materials with a wavelength of 1.5–1.6 μm [4,5]. The $LnAl_3(BO_3)_4$ crystals with the Ln^{3+} = Tb, Ho, Er, Tm exhibit a magnetoelectric effect ($HoAl_3(BO_3)_4$ is a leader among these compounds) (for example, [2,7]). The $LnAl_3(BO_3)_4$ compounds with the Ln^{3+} = Gd, Eu, Tb, Ho, Pr, Sm are used as phosphors (for example, [8,9]): the $YAl_3(BO_3)_4$ crystals doped with the Eu^{3+} and Tb^{3+} ions is an environmentally friendly material for a white LED, having high intensity and luminescence power and low cost [10]; those doped with the Tm^{3+} and Dy^{3+} are able to be tuned from blue through white and ultimately to yellow emission colors and are attractive candidates for general illumination [11]; those doped with the Sm^{3+} ions, as well as $GdAl_3(BO_3)_4:Sm^{3+}$, can be used as promising materials for orange-red lasers [8]. The simultaneous generation of three basal red–green–blue colors is obtained from a lone $GdAl_3(BO_3)_4:Nd^{3+}$ bi-functional laser and optical nonlinear crystal [12], which is also of interest for the development of high-quality and bright displays.

Rare-earth gallium borates $LnM_3(BO_3)_4$ with the M^{3+} = Ga are poorly studied. The $LnGa_3(BO_3)_4$ crystals, in particular $LnGa_3(BO_3)_4:Tb^{3+}$ with the Ln^{3+} = Y or Gd, are prominent luminescent materials with plasma-discharge conditions, converting a vacuum ultraviolet radiation into a visible light that can be used in high-performance plasma display panels and television devices [13]. In addition, gallium borates can be considered as promising materials not only for luminescent and laser applications, but also for use in spintronics: a large magnetoelectric effect was found in the $HoGa_3(BO_3)_4$ [14].

Rare-earth borates $LnM_3(BO_3)_4$ with the magnetic ions M^{3+} = Fe or Cr, which are characterized by the presence of two interacting magnetic subsystems (3*d*- and 4*f*- ions) in the crystal structure, are even less studied. A long-range antiferromagnetic spin order of the Cr^{3+} subsystem is observed in the $NdCr_3(BO_3)_4$ single crystals at T_N = 8 K [15]. A phase transition from paramagnetic to antiferromagnetic state was found in the $EuCr_3(BO_3)_4$ crystals at $T_N \approx 9$ K [16]. In addition, a magnetic phase transition is also observed in the $SmCr_3(BO_3)_4$ at $T_c = 5 \pm 1$ K [17]. In a number of rare-earth ferroborates, a significant magnetoelectric effect was found [18,19], which makes it possible to attribute them to a new class of multiferroics [18]. Maximum magnetoelectric and magnetodielectric effects were recorded for the $NdFe_3(BO_3)_4$ and $SmFe_3(BO_3)_4$ crystals [19,20]. Such materials can be used as magnetoelectric sensors, memory elements, magnetic switches, spintronics devices, high-speed radiation-resistant MRAM memory, etc.

Rare-earth scandium borates $LnM_3(BO_3)_4$ with the M^{3+} = Sc demonstrate an anomalously low luminescence concentration quenching, which is caused by the large distance between the nearest

Ln ions (~6 Å), and, hence, these compounds are promising high-efficient optical media that can be used in photonics, in particular, to create a diode-pumped compact lasers covering various optical spectral regions [21,22]. Currently, the best materials for medium-power solid-state miniature lasers are undoped and Cr^{3+}-doped $NdSc_3(BO_3)_4$ [22,23]. They are characterized by a high efficiency of laser transitions, on the one hand, and detuning of almost all cross-relaxation transitions, on the other. In addition, the $NdSc_3(BO_3)_4$ crystals have a high non-linear dielectric susceptibility. Hence, in these crystals, it is possible to convert an infrared radiation of the neodymium laser into a visible one along a certain crystal orientation relative to the propagation of laser radiation. A miniature laser emitted in a green spectral region can be created based on the $NdSc_3(BO_3)_4$.

As can be seen from the short literature review on properties and possible areas of application of the huntite-family borates, these materials are characterized by a combination of different functional properties in one compound. In turn, physical and chemical properties are due to a real composition and crystal structure, influenced by initial composition (the type of rare-earth metal and the nature of *M* metal), synthesis conditions, and external effects.

Determination of structure of huntite-family rare-earth borates becomes very important, since a possible structural transition from one space group to another can be accompanied by a loss (or acquisition) of the center of symmetry and results in an acquisition (or loss) of nonlinear optical and magnetoelectric properties. Sardar et al. [24] found that the introduction of 5 at % Nd^{3+} ions $(2.3 \times 10^{20} \text{ cm}^{-3}$; activation) into the $LaSc_3(BO_3)_4$ crystal leads to a change in symmetry from the space group $C2/c$ to $C2$, i.e., to a transition from a centrosymmetric to a non-centrosymmetric structure. In addition, it was shown that the introduction of Nd^{3+} ions in a concentration more than or equal to 50 at % to the $LaSc_3(BO_3)_4$ crystal (solid solution) results in the space group transition from $C2/c$ to $R32$ [25]. A comprehensive study of solid solutions in the system $NdCr_3(BO_3)_4$ (sp. gr. $C2/c$)–$GdCr_3(BO_3)_4$ (sp. gr. $R32$) using spectroscopic methods showed that the $Nd_xGd_{1-x}Cr_3(BO_3)_4$ borates with x < 0.2 have essentially trigonal non-centrosymmetric structure (sp. gr. $R32$), and already in the case of 20% Nd concentration in the crystals, an additional large content of the monoclinic phase (sp. gr. $C2/c$) is observed [26].

Knowledge of a precise real composition of crystals is no less important. Bulk crystals, usually obtained by the most technological melt methods, can have a homogeneous composition only for compounds with the congruent melting (**CM**). However, an activation of crystal or a synthesis of mixed crystals (solid solutions) leads to a deviation of a real composition from the CM one. In addition, a real composition of grown crystal, taking into account compositions of all crystallographic sites, usually differ from that of initial charge (melt). However, in the overwhelming majority of cases, observed physical properties of materials are 'attributed' to initial charge composition that leads to incorrect conclusions about correlations in the fundamental triad 'composition-structure-properties'.

Thus, to improve properties of huntite-family single crystals and expand the scope of their application as well as to synthesize and create materials with a required combination of operating parameters, it is necessary to know a precise real crystal composition, structural effects, and crystallochemical limits of existence (stability limits) of a compound or solid solution with a specific symmetry. This is the motivating force for the research, the results of which are reported on here.

Currently, various approaches are developed to avoid time-consuming searches for materials with a specific crystal structure and desired physical parameters and to optimize and simplify ways of their obtaining. One of the most effective techniques is a type of crystal engineering which is shown here for the huntite-family compounds. In this approach, the 'composition-structure-property' correlations for a specific class of materials as well as possible structural types with chemical element sets formed crystal structure, coordination environment of atoms in the structure, nature of chemical bond between different atomic groups, etc. are analyzed. When forecasting new structures or describing the known ones, it is necessary to take into account crystallochemical characteristics of atoms or ions in the structure, namely, a radius, an electronegativity, a formal charge, and hence a coordination number and bonds between the components. An approach described is one of the main driving forces in the applied crystallochemistry—a section of materials science, which combines knowledges from fundamental crystallochemistry and specialists in other scientific fields dealing with materials. One of the main tasks of applied crystallochemistry is an investigation of functional correlations of the form of $P = P(X)$, where X is a material and P is a property. In this case, properties are considered depending on the crystallochemical individuality of the components in a number of related compounds. A computer design of such structures involves a generation of polyhedra, a search for possible packages with the following determination of all required structural parameters.

2. Materials and Methods

In this review, single crystals are predominantly described and a preference is given to X-ray diffraction research techniques—a powerful tool to determine fundamental characteristics of an object, in particular, a real composition and structure, with a high degree of accuracy and reliability and to reasonably relate functional properties to a composition and structure depending on the prehistory of crystals (composition of initial charge, growth and post-growth treatment conditions). A crystallochemical approach based on the knowledge of basic laws of structure formation and correlations between real composition, structure, growth conditions, and physical properties of materials, makes it possible to optimize a search for new promising compounds and to improve functional properties of the known ones. The use of information technologies applied crystallochemical models for a subsequent creation of mathematical models and specialized programs on their basis accelerates a transition from theory to practice, and further to a technology to grow material with controlled properties.

The $LnM_3(BO_3)_4$ crystals, where Ln^{3+} = La–Lu, Y, M^{3+} = Al, Fe, Ga, Cr, melt incongruently and are usually obtained by the flux method from the high-temperature solutions both via spontaneous crystallization and using a seed, which is described in detail in [21,27–29]. The use of the flux method, as the most suitable for growing such crystals, has a significant drawback—an incorporation of flux components into the growing crystal. The $K_2Mo_3O_{10}$–B_2O_3 mixed flux is usually applied to grow huntite-family borates [21,27,30], however, the Mo ions easily incorporate into a crystal, which leads to an appearance of a near ultraviolet absorption band that inhibits the use of these nonlinear optical crystals at short wavelength [21,27,31]. In addition, crystals have low growth rate and small sizes (1–20 mm), which leads to an impossibility of obtaining bulk samples of good optical quality suitable for further use. In [21,32,33] only, the relatively large $LnM_3(BO_3)_4$ crystals, in particular, $YAl_3(BO_3)_4$ and $GdAl_3(BO_3)_4$, up to 45 mm in size have been synthesized by the top-seeded solution growth (**TSSG**) method.

In addition, it should be noted that the polycrystalline gallium borates with the Ln = La, Nd, Sm, Gd, Ho, Y, Er, and Yb were sintered in Pt crucible in air atmosphere at T = 575–1050 °C from pellets made from Ln_2O_3, Ga_2O_3, and B_2O_3 oxide powders [34]. As a result, the dominant metaborate $Ln(BO_2)_3$ for the Ln = La and Nd, huntite $LnGa_3(BO_3)_4$ for the Ln = Sm, Gd, Ho, Y and Er, the new dolomite $YbGa(BO_3)_2$, the intermediate $LnBO_3$ and $GaBO_3$ phases were identified by X-ray powder diffraction measurements.

A significant advantage of the $LnSc_3(BO_3)_4$ scandium borates is a possibility of synthesis of large-sized single crystals using the Czochralski technique (Ln = La, Ce, Pr, Nd; in particular, [35]). Wherein, the degree of congruence of $LnSc_3(BO_3)_4$ decreases from La to Nd [35], and compounds with the Ln = Y and Gd melt incongruently [36,37]. Durmanov et al. [33] developed the modified heating assembly served to effectively control both the process of possible condensation of B_2O_3 vapors on the surface of the growing crystal and thermal gradients over the crucible and in the melt. It guaranteed synthesis of high-quality optical $LaSc_3(BO_3)_4$ single crystals (both pure and doped with the Nd, Yb, Er/Yb, Er/Yb/Cr, Pr), $CeSc_3(BO_3)_4$, $PrSc_3(BO_3)_4$ (both pure and doped with the Nd), $NdSc_3(BO_3)_4$, as well as solid solutions, in particular, those with the general chemical compositions $(Ce,Nd)Sc_3(BO_3)_4$, $(Ce,Gd)Sc_3(BO_3)_4$, $(Nd,Gd)Sc_3(BO_3)_4$, $(Ce,Nd,Gd)Sc_3(BO_3)_4$, $(Ce,Y)Sc_3(BO_3)_4$, $Ce(Lu,Sc)_3(BO_3)_4$, $(Ce,Nd)(Lu,Sc)_3(BO_3)_4$.

Single crystals with the general composition $LnSc_3(BO_3)_4$ (Ln = La, Ce, Pr, Nd, Tb), in particular, with the initial charge compositions $LaSc_3(BO_3)_4$, $CeSc_3(BO_3)_4$, $Pr_{1.1}Sc_{2.9}(BO_3)_4$, $Pr_{1.25}Sc_{2.75}(BO_3)_4$, $NdSc_3(BO_3)_4$, $Nd_{1.25}Sc_{2.75}(BO_3)_4$, $TbSc_3(BO_3)_4$ and single-crystal solid solutions with the initial compositions $(La,Nd)Sc_3(BO_3)_4$, $(Ce,Nd)_{1+x}Sc_{3-x}(BO_3)_4$, $(La,Pr)Sc_3(BO_3)_4$, $La(Sc,Yb)Sc_3(BO_3)_4$, $La(Sc,Er)Sc_3(BO_3)_4$, $La(Sc,Er,Yb)Sc_3(BO_3)_4$, $(Ce,Gd)Sc_3(BO_3)_4$, $(Ce,Gd)_{1+x}Sc_{3-x}(BO_3)_4$, $(Nd,Gd)_{1+x}Sc_{3-x}(BO_3)_4$, $(Ce,Nd,Gd)Sc_3(BO_3)_4$, $(Ce,Y)_{1+x}Sc_{3-x}(BO_3)_4$, $(Ce,Lu)_{1+x}Sc_{3-x}(BO_3)_4$, $(Ce,Nd,Gd)_{1+x}Sc_{3-x}(BO_3)_4$ $(Ce,Nd,Lu)_{1+x}Sc_{3-x}(BO_3)_4$, described in this work, having averaged diameter of 15–25 mm and length of 30–150 mm were grown by the Czochralski technique in Ir crucibles at pulling rate 1–3 mm/h and seed rotation 8–12 rpm. An Ir rod of 2 mm in diameter was initially used as a seed. Oriented single-crystal seeds were cut from the crystals grown on the Ir rod. The seed was oriented so that its optical axis coincided with the pulling axis (within a few degrees). The methodology for Czochralski growth technique used to synthesis $LnSc_3(BO_3)_4$ crystals (Ln = La, Ce, Pr, Nd, Tb) as well as single-crystal solid solutions is given in [38,39].

According to the literature data, a crystallization of $LnM_3(BO_3)_4$ compounds (Ln^{3+} = La–Lu, Y; M^{3+} = Al, Fe, Ga, Cr) in a specific space group was found using different techniques, namely, diffraction methods (**D**), infrared spectroscopy (**IR**), absorption spectroscopy (**AS**), transmission spectroscopy (**TS**), Raman spectroscopy (**R**); in addition, temperatures of structural phase transitions between forms with different space groups were determined using specific heat measurements (**SH**) and differential thermal analysis (**DTA**) [1,15,16,26,27,34–37,40–89] (Table 1). In several works [36,49], the structures of the compounds were declared solely on the basis of a comparison of the experimental diffraction patterns with those given in the structural databases without any detailed structural analysis. In the overwhelming number of cases, the crystal structures (first of all, coordinates of atoms) of the samples were refined by the full-profile Rietveld method on polycrystalline samples obtained by the solid-state reaction ($LnFe_3(BO_3)_4$ with the Ln = La, Ce, Pr, Nd, Sm, Eu, Gd, Tb, Dy, (Y), Ho; $LnGa_3(BO_3)_4$ with the Ln = Sm, Gd, Y, Ho, Er) [34,53] or on single crystals synthesized by the flux method and ground to a powder ($LnAl_3(BO_3)_4$ with the Ln = Nd, Sm, Eu, Gd, Tb, Dy, (Y), Ho, Er, Yb; $LnFe_3(BO_3)_4$ with the Ln = Ce, Pr, Nd, Sm, Eu, Gd, Tb, Dy, (Y), Ho; $LnGa_3(BO_3)_4$ with the Ln = Gd) [1,54,56,59,61,66,68]. Assignment of a series of rare-earth orthoborates ($LnAl_3(BO_3)_4$ with the Ln = Nd, Eu, Gd, Tb, Dy, (Y), Ho, Er, Tm, Yb; $LnFe_3(BO_3)_4$ with the Ln = Eu, Gd, Y; $LnCr_3(BO_3)_4$ with the Ln = Sm, Eu, Gd, Tb, Dy; $LnGa_3(BO_3)_4$ with the Ln = Nd, Eu, Gd, (Y), Ho) to the space group $R32$ is performed on powdered single crystals obtained by the flux method using IR spectroscopy coupled with the group-theoretical analysis [26,42,45,51,70,71,73], temperature-dependent high-resolution optical absorption Fourier spectroscopy and Raman spectroscopy ($LnFe_3(BO_3)_4$ with the Ln = Pr, Nd) [57,58,60].

Crystal structures of single crystals were refined within the framework of the huntite structure (space group $R32$) for the $LnAl_3(BO_3)_4$ (Ln = Nd, Sm, Eu, Gd, (Y), Tm, Yb), $LnFe_3(BO_3)_4$ (Ln = La, Nd, Eu, Gd, Er), $LnGa_3(BO_3)_4$ (Ln = Nd, Eu, Ho), $LnSc_3(BO_3)_4$ (Ln = La, Ce, Pr, Sm, Eu) by the X-ray diffraction (**XRD**) analysis (Table 1). It should be noted that in some literature sources (in particular, in [27]) it was not indicated on which samples, single-crystal or polycrystalline, a structural analysis has been performed; in [15,36,37,80], any methodology of structural studies is absent, the method of investigation, X-ray diffraction, being indicated only. Neutron diffraction study performed on powdered samples with the initial compositions $YAl_3(BO_3)_4$, $Y_{0.88}Er_{0.12}Al_3(BO_3)_4$, $Y_{0.5}Er_{0.5}Al_3(BO_3)_4$, $Y_{0.5}Yb_{0.5}Al_3(BO_3)_4$, and $Y_{0.84}Er_{0.01}Yb_{0.15}Al_3(BO_3)_4$, grown by the topseeded high temperature solution method, crystallized in the space group $R32$ [90]. Due to the fact that the neutron scattering amplitudes both for Er (b = 7.79 fm) and for Yb (b = 12.433 fm) are much greater than that for Al (b = 3.449 fm), according to the Rietveld calculations, it can be stated that the Er^{3+} and Yb^{3+} ions occupy the Y^{3+} sites. For the above-mentioned compounds, positional parameters have been refined only [90].

Table 1. Space groups for the compounds with the general composition $LnM_3(BO_3)_4$ with the M^{3+} = Al, Fe, Cr, Ga, Sc (according to the literature data).

Ln	Space groups for the $LnM_3(BO_3)_4$ [1]				
	M = Al	M = Fe	M = Cr [2]	M = Ga	M = Sc
La	Orthorhombic symmetry: D/P [40]	R32: D/? [27], D/P [53,54], D/S [55]	C2/c: IR/S (70:30, 1040–1050 °C) [70], IR/S (1.5:1; 2.3:1) [71]		R32: D/S [74,75] C2/c: D/S (or C2) [76], D/S [77], D/P [78], Cc: D/S [79]
Ce		R32: D/P [73]			R32: ? [80], D/S [81] C2/c: D/S [76,77,83–86] Refined composition: CeSc₃(BO3)4 [82]
Pr	R32: D/? [27] C2/c: D/? [27], D/S [41], IR/P [42] C2: D/? [27]	R32: D/P [73], D/P (1.5, 300 K) [56], AS/S [57,58]	C2/c: IR/S (50:50, 900–950 °C, 1040–1050 °C) [70], IR/S (1:1; 1.5:1; 2.3:1) [71]		R32: ? [80], D/S [88] C2/c (or C2): D/S [15,83] P321: D/S (40% reflections R32) [82] Refined compositions: $(Pr_{0.919}Sc_{0.081(4)})Sc_3(BO_3)_4$ [82], $(Pr_{0.924}Sc_{0.076(4)})Sc_3(BO_3)_4$ [82]
Nd	R32: D/P [1], D/? [27], D/S [43,44], IR/P [42,45] C2/c: D/S [46,47], IR/P [42,45], D/P [48], ? [49] C2: D/P [48], ? [49]	R32: D/? [27], D/P [53,54,59], D/S [7], R/S [61], AS/S [57,58] C2/c: IR/P [45]	R32: D/? [27], ? [15] C2/c: IR/S (50:50, 900–950 °C; 70:30, 1040–1050 °C) [70], IR/S (1:1; 2.3:1) [71], IR/S [2,64]	R32: D/S [47], D/? [2], IR/P [45]	R32: ? [80],], D/S [7], D/S [88] P321 (or P3): D/S [9,87,89] P321: D/S (40% reflections R32) [82] Refined compositions: NdSc₃(BO3)4 [82,89], $(Nd_{0.910}Sc_{0.090(20)})Sc_3(BO_3)_4$ [82]
Sm	R32: D/P [1], D/? [27], D/S [50] C2/c: D/? [27], IR/P [42]		R32: D/? [27], IR/S (1:1) [71], IR/S (50:50, 900–950 °C) [70], IR/S [2] C2/c + C2/c fragments: IR/S (1.5:1) [71], C2/c + R32: IR/S (2.3:1) [71]	R32: D/? [2], D/P [84]	R32: ?], D/S [88]
Eu	R32: D/P [1], D/? [27], D/S [50], IR/P [42,51] C2/c: D/? [27] C2: D/? [41]	R32: D/? [27], D/S [60], D/P [53,54], IR/P [45] R32 (HT) and P3₁21 (LT): SH/S, T_s = 88 K [15]; AS/S, T_s = 58 K [57,58]; TS/S, T_s = 84 K [62]	R32: D/? [27], D/P [16] IR/S (50:50, 900–950 °C, 70:30, 1040–1050 °C) [70] IR/S (1:1; 2.3:1) [71], IR/S [41]	R32: D/? [2], D/S [74] IR/P [51,74]	R32: D/S [88]
Gd	R32: D/P [1], D/? [27], D/S [50], D/P [27], IR/P [42] C2: D/S [46,47]	R32: D/? [27], D/P [34,54], D/S (297 K) [63], IR/P [45] P3₁,21: D/S (90 K) [63] R32 (HT) and P3₁21 (LT): SH/S, T_s = 174 K [15]; R/S, T_s = 155 K [60]; AS/S, T_s = 133–156 K [57,58,61]; IR/S, T_s = 143 K [65]	R32: D/P [1], D/? [27], IR/S (50:50, 900–950 °C; 70:30, 1040–1050 °C) [70] IR/S (1:1; 1.5:1; 2.3:1) [71], IR/S [26,45]	R32: D/? [2], D/P [34,54], IR/P [43]	R32: ? [36,37]

Table 1. *Cont.*

Ln	Space groups for the $LnM_3(BO_3)_4$ [1]				
	M = Al	M = Fe	M = Cr [2]	M = Ga	M = Sc
Tb	**R32**: D/P [1], D/? [27], IR/P [42]; **C2/c**: D/S [41]	**R32**: D/? [27], D/P [53,54], D/P (200 K, 300 K) [66]; **P3₁21**: D/P (2, 30, 40, 100 K) [66]; **R32 (HT) and P3₁21 (LT**: SH/S, T_s = 241 K [53]; R/S, T_s = 198 K [64]; AS/S, T_s = 198 K [7,58]; IR/S, T_s = 192 K [64]; IR/S, T_s = 200 K [65]	**R32**: D/? [27], IR/S (50:50, 900–950 °C) [70]; **R32 + C2/c fragments**: IR/S (1:1) [71]; **C2/c + R32 fragments**: IR/S (1.5:1) [71]; **C2/c**: IR/S (70:30, 1040–1050 °C) [70], IR/S (2.3:1) [71]	**R32**: D/? [27]	**R3̄ or R3-Calcite-type structure**: D/S [57,58] *Refined composition*: (Tb₀.₂₅Sc₀.₇₅)BO₃ [83]
Dy	**R32**: D/P [1], D/? [27], IR/P [42]	**R32**: D/? [27], D/P [53,54], **P3₁21**: D/P (1.5, 50, 300 K) [67]; **R32 (HT) and P3₁21 (LT**: SH/S, T_s = 340 K [53]	**R32**: IR/S (50:50, 900–950 °C) [70], IR/S (1.5:1) [71]; **R32 + C2/c fragments**: IR/S (1:1) [71]; **C2/c**: IR/S (70:30, 1040–1050 °C) [70], IR/S (2.3:1) [71]	**R32**: D/? [27]	
(Y)	**R32**: D/P [1], D/S [45,47], D/? [27], IR/P [42,46]	**R32**: D/? [27], D/P [53,54], IR/P [43], D/P (520 K) [68]; **P3₁21**: D/P (2, 50, 295 K) [68]; **R32 (HT) and P3₁21 (LT**: DTA/S, T_s = 445 K [53]; R/S, T_s = 350 K [60]; AS/S, T_s = 350 K [7,58]		**R32**: D/? [27], D/P [34], IR/P [43]	**R32**: ? [96,97]
Ho	**R32**: D/P [1], D/? [27], IR/P [42]; **C2/c**: D/S [41]	**R32**: D/? [27], D/P [53,54], D/P (520 K) [68]; **P3₁21**: D/P (2, 50, 295 K) [68]; **R32 (HT) and P3₁21 (LT**: DTA/S, T_s = 427 K [53]	**R32**: D/? [27]; **R32 + C2/c fragments**: IR/S (1.5:1) [7,1]; **C2/c**: IR/S (70:30, 1040–1050 °C) [70], IR/S (2.3:1) [71]	**R32**: D/? [27], D/S [73], D/P [34], IR/P [73]	
Er	**R32**: D/P [1], D/? [27], IR/P [42]	**R32**: D/? [27], D/S [69], **P3₁21**: D/P (1.5, 300 K) [56]; **R32 (HT) and P3₁21 (LT**: R/S, T_s = 340 K [6]; AS/S, T_s = 340 K [7,58]	**R32 + C2/c fragments**: IR/S (1:1) [71]	**R32**: D/? [27], D/P [34]	
Tm	**R32**: D/? [27], D/S [41], IR/P [42]	**R32**: D/? [27]			
Yb	**R32**: D/P [1], D/? [27], D/S [41], IR/P [42]	**R32**: D/? [27]	**R32**: D/? [27]	**R32**: D/? [27]; **R3-Dolomite type structure**: D/P [34]	
Lu	**R32**: D/? [27]				

¹ P—powder sample, S—single crystal sample, D—diffraction techniques, IR—infrared spectroscopy, R—Raman spectroscopy, TS—transmission spectroscopy, AS—absorption spectroscopy, SH—specific heat measurements, DTA—differential thermal analysis, HT—high-temperature phase, LT—low-temperature phase, T_s—phase transition temperature; A question mark (?) indicates a lack of data on the type of material under investigation or/ and applied technique in the literature source. ² The borate:solvent ratios in the batch and the temperature ranges are given in parentheses.

Resuts of XRD study of the Czochralski-grown single crystals and solid solutions ("Enraf-Nonius" CAD-4 single-crystal diffractometer; room temperature; AgK_α, MoK_α or CuK_α; size, ~ 0.1 × 0.1 × 0.1 mm^3) with the initial compositions $LaSc_3(BO_3)_4$, $CeSc_3(BO_3)_4$, $Pr_{1.1}Sc_{2.9}(BO_3)_4$, $Pr_{1.25}Sc_{2.75}(BO_3)_4$, $NdSc_3(BO_3)_4$, $Nd_{1.25}Sc_{2.75}(BO_3)_4$, $TbSc_3(BO_3)_4$ and $(La,Nd)Sc_3(BO_3)_4$, $(Ce,Nd)_{1+x}Sc_{3-x}(BO_3)_4$, $(La,Pr)Sc_3(BO_3)_4$, $La(Sc,Yb)Sc_3(BO_3)_4$, $La(Sc,Er)Sc_3(BO_3)_4$, $La(Sc,Er,Yb)Sc_3(BO_3)_4$, $(Ce,Gd)Sc_3(BO_3)_4$, $(Ce,Gd)_{1+x}Sc_{3-x}(BO_3)_4$, $(Nd,Gd)_{1+x}Sc_{3-x}(BO_3)_4$, $(Ce,Nd,Gd)Sc_3(BO_3)_4$, $(Ce,Y)_{1+x}Sc_{3-x}(BO_3)_4$, $(Ce,Lu)_{1+x}Sc_{3-x}(BO_3)_4$, $(Ce,Nd,Gd)_{1+x}Sc_{3-x}(BO_3)_4$ $(Ce,Nd,Lu)_{1+x}Sc_{3-x}(BO_3)_4$, performed by our scientific group, are given in [35,76,82,83,85,89,91] and [35,84,86,87,91], respectively, and in the present work. To reduce an error associated with an absorption, the XRD data were collected over the entire Ewald sphere. The unit cell parameters are determined by an auto-indexing of the most intense 25 reflections. Furthermore, a detailed analysis of diffraction reflections, including low-intensity ones, was carried out to find a possible superstructure either with the multiple-increased unit cell parameters and another symmetry or with another symmetry only. In the case of a small number of reflections that do not obey the extinction rules of the chosen space group, these reflections were not taken into account when refining a crystal structure. In case of a large number of 'forbidden' reflections, a crystal structure was solved by direct methods or/and the Paterson method taking into account all diffraction reflections. The preliminary XRD data processing was carried out using the WinGX pack [92] with a correction for absorption. The atomic coordinates, anisotropic displacement parameters of all atoms, and occupancies of all the sites (except for the B and O ones) were refined using the SHELXL2013 software package [93], taking into account the atomic scattering curves for neutral atoms. The structural parameters were refined in several steps: initially, the coordinates of 'heavy' atoms (*Ln* and Sc), and then those of 'light' atoms (O and then B) were refined together with the atom displacement parameters in isotropic and then anisotropic approximations; finally, the occupancies of the *Ln* and Sc sites were refined step by step.

In the review, for the visualization and comparison of crystal structures (ball-and-stick and polyhedral models) and individual polyhedra as well as for the calculation of structural parameters (all interatomic distances, bond angles, etc.) of the huntite-family compounds and solid solutions with different symmetry, the improved and augmented computer program for the investigation of the dynamics of changes in structural parameters of compounds with different symmetry has been applied [94]. Theoretical diffraction patterns have been created using the DIAMOND [95] and specialized software developed for diffraction pattern indexing taking into account a selection of background level for rhombohedral and hexagonal cells of the huntite structure and a refinement of unit cell parameters using different sets of diffraction reflections: CPU, Intel core i3; RAM, at least 4 GB; Code, C#; OS: Windows 7 with the installed Microsoft.NET Framework 4.0 or higher; Size: 32 768 b.

3. Results

3.1. $LnM_3(BO_3)_4$ (M = Al, Fe, Cr, Ga, Sc) Compounds

All the known literature data on symmetry of the huntite-family compounds, determined mainly by diffraction and spectroscopic methods on both polycrystalline and single-crystal samples, are systematized and given in Table 1 and Figure 1. It can be noted that almost all compounds having the general composition $LnM_3(BO_3)_4$ with the M^{3+} = Al, Fe, Cr, Ga, Sc have a modification with the huntite structure (space group $R32$).

The situation looks different if only the results of X-ray study of single crystals are taken into account (Figure 2): the compounds with the general composition $LnM_3(BO_3)_4$ with the M^{3+} = Al и Sc are most fully represented; a modification with the huntite structure (space group $R32$) prevails.

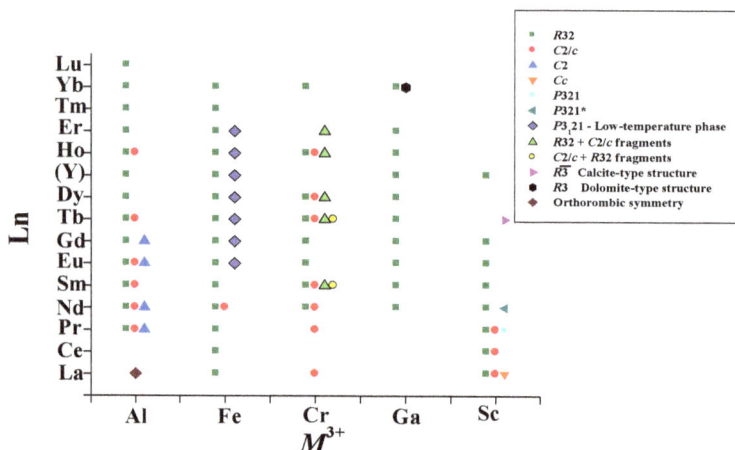

Figure 1. Space groups for the compounds with the general composition $LnM_3(BO_3)_4$ with the M^{3+} = Al, Fe, Cr, Ga, Sc (according to the literature data given in Table 1).

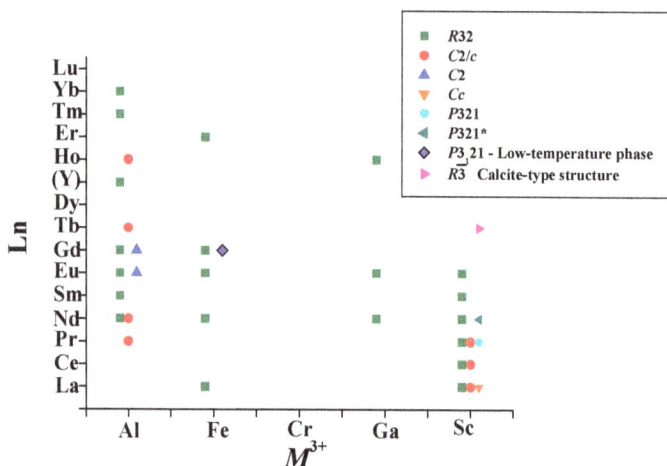

Figure 2. Space groups for the single crystals with the general composition $LnM_3(BO_3)_4$ with the M^{3+} = Al, Fe, Ga, Sc (according to the literature data on X-ray diffraction experiments given in Table 1).

In the crystal structure of the huntite $CaMg_3(CO_3)_4$ (space group $R32$, a = 9.5027(6), c = 7.8212(6) Å, Z = 3) (Figure 3), the Ca^{2+} ion (r_{Ca}^{VI} = 1.00 Å according to the Shannon system [96]) is located in the center of a distorted prism with the CN Ca = 6 (CN, coordination number). The upper triangular face of prism is rotated with respect to the lower one by an angle φ = 9.596° (φ = 0° in the regular trigonal prism), the Ca–O interatomic distances being the same. The Mg^{2+} ion (r_{Mg}^{VI} = 0.72 Å) is located in the center of a distorted octahedron with three different Mg–O interatomic distances (CN Mg = 2 + 2 + 2). Crystallochemically-different B1 and B2 ions occupy the centers of isosceles (CN B1 = 1 + 2) and equilateral (CN B2 = 3) triangles, respectively. The MgO_6 octahedra are joined by the edges and form twisted chains extended parallel to the c axis (the 3_1 axis). The B2 atoms are located at the two-fold axes in triangles between the chains from the MgO_6 octahedra, forming a 'spiral staircase' around the 3_2 axes. Different chains are connected by the CaO_6 trigonal prisms and BO_3 triangles, where each individual CaO_6 and BO_3 group connects three chains [55,65]. Figure 3 shows the XY and XZ projections of huntite-type unit cell of $LnM_3(BO_3)_4$ (space group $R32$) as a ball-and-stick model (Figure 3a,b) and polyhedra (Figure 3c,d) as well as fragments of the structure, including the main coordination polyhedra (Figure 3e,f), and individual coordination polyhedra for all crystallochemically-different atoms in the crystal structure (Figure 3g–i).

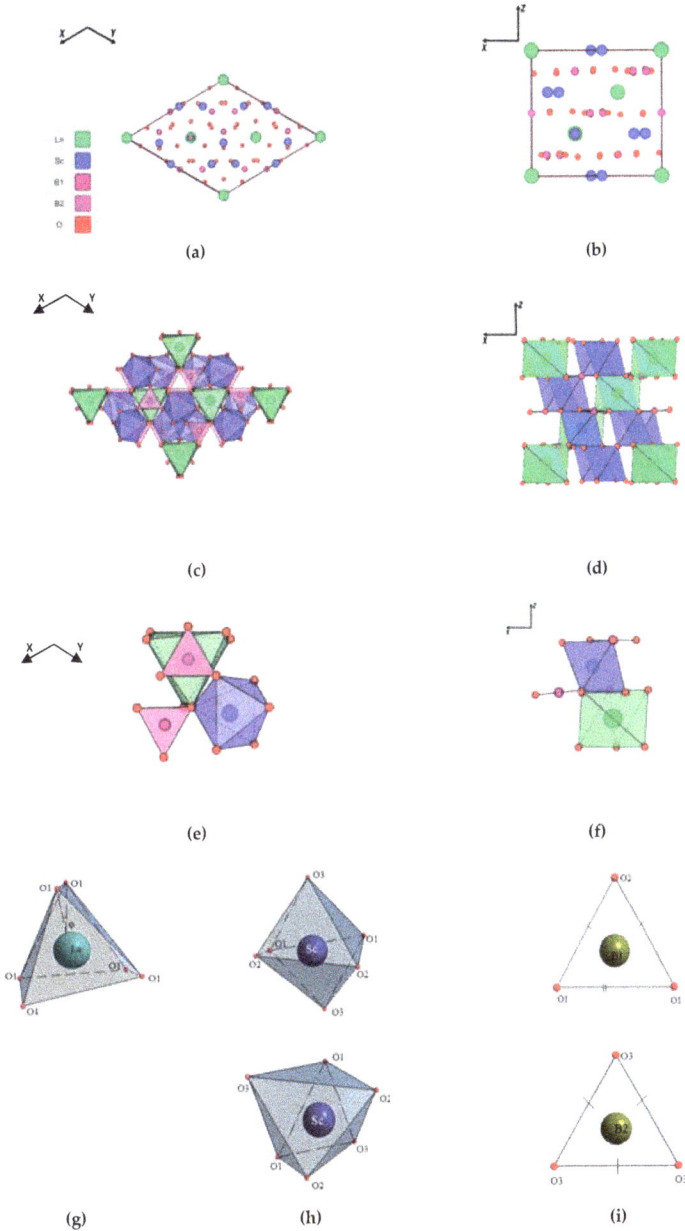

Figure 3. The unit cell of the $LnM_3(BO_3)_4$ structure (space group $R32$) projected onto the (**a**) XY and (**b**) XZ planes; Combination of the coordination polyhedra projected onto the (**c**) XY and (**d**) XZ planes; Combination of selected coordination polyhedra projected onto the (**e**) XY and (**f**) XZ planes; Coordination polyhedra for the (**g**) Ln, (**h**) M, (**i**) B1 and B2.

A topological correspondence of the $Ca^{2+}Mg^{2+}_3(CO_3)^{4-}_4$ and $Ln^{3+}M^{3+}_3(BO_3)^{3-}_4$ formulas, a similar coordination environment of the C^{4+} and B^{3+} ions, a possibility of the compensation of system electroneutrality, and a presence of the Ca^{2+} and Ln^{3+}, Mg^{2+} and Sc^{3+}, Mg^{2+} and M^{3+} = Al, Ga, Fe ions in the Goldschmidt–Fersman diagonal series should lead to the isostructurality of the $Ca^{2+}Mg^{2+}_3(CO_3)^{4-}_4$ and $Ln^{3+}M^{3+}_3(BO_3)^{3-}_4$ compounds (more precisely, these compounds are isotype, as evidenced by the lack of similarity in structures, typical of isostructural compounds [91]) despite the fundamental difference in the crystallochemical properties (dimensions, electronegativity values, formal charges).

It should be noted that the Cr^{3+} ions are surrounded by six O atoms, forming the regular CrO_6 octahedra (CN Cr = 6) in the crystal structures of oxides due to the electronic structure (non-binding configuration is symmetric to the octahedral field of ligands - d_ε^3), unlike distorted MO_6 octahedra in the huntite structure (another example confirmed that the $Ca^{2+}Mg^{2+}_3(CO_3)^{4-}_4$ and $Ln^{3+}Cr^{3+}_3(BO_3)^{3-}_4$ compounds are isotype). Hence, in the structure of the activated $La^{3+}Sc^{3+}_3(BO_3)^{3-}_4$:$Cr^{3+}$ crystal, the symmetry of the ScO_6 polyhedra, which include Cr^{3+} ions, as well as that of the whole crystal changes. This is confirmed by the XRD analysis of the $La^{3+}Sc^{3+}_3(BO_3)^{3-}_4$ and $La^{3+}Sc^{3+}_3(BO_3)^{3-}_4$:Cr single crystals, grown by the Czochralski method, with the monoclinic (space group $C2/c$) and triclinic (space group $P1$ or $P\bar{1}$) symmetry, respectively [76]. In the latter case, quite a lot of reflections with the $I < 3\sigma(I)$ are not indexed in the monoclinic syngony, and taking into account these reflections, the refined unit cell parameters of the $LaSc_3(BO_3)_4$:Cr were found to be a = 7.7356(4), b = 9.8533(8), c = 12.0606(8) Å, α = 89.981(6), β = 105.437(5), γ = 90.045(6)°, in contrast to those of the $LaSc_3(BO_3)_4$, a = 7.727(1), b = 9.840(1), c = 12.046(3) Å, β = 105.42(2)°.

In the $CaCO_3$-$MgCO_3$ system, the $CaCO_3$ ($a \approx 4.99$, $c \approx 17.08$ Å; space group $R\bar{3}c$, Z = 6) and $CaMg(CO_3)_2$ ($a \approx 4.80$, $c \approx 16.00$ Å; space group $R\bar{3}$, Z = 3) compounds with calcite and dolomite structures, respectively, are known. The Ca^{2+} and Mg^{2+} ions occupy regular and distorted octahedral sites in the calcite and dolomite structures, respectively, with an ordered arrangement of the Ca^{2+} and Mg^{2+} ions along the 3-fold axis, which leads to a decrease in the symmetry of $CaMg(CO_3)_2$ compared to the $CaCO_3$. Based on the transformed compositions ($CaCO_3 \equiv Ca_4(CO_3)_4$, $CaMg(CO_3)_2 \equiv Ca_2Mg_2(CO_3)_4$) and addiction of Ln^{3+} ions to predominantly trigonal-prismatic (Ln = La–Gd) or octahedral (Ln = Tb–Lu; as well as M^{3+} ions) coordination [35], it is possible that compounds with the initial composition $Ln^{3+}M^{3+}_3(BO_3)^{3-}_4$ can have dolomite-like and calcite-like structures with an ordered arrangement of Ln^{3+} and M^{3+} ions over the octahedral sites both separately (full positional ordering) and jointly (partial positional ordering). This can be expected, for example, for the $Ln^{3+}M^{3+}_3(BO_3)^{3-}_4$ with the Ln = Tm (r_{Tm}^{VI} = 0.88 Å) or Yb (r_{Yb}^{VI} = 0.87 Å) in combination with the M^{3+} = Cr (r_{Cr}^{VI} = 0.615 Å) or Ga (r_{Ga}^{VI} = 0.620 Å), for which Δr_{Ln-M} = ~0.25 Å, forming the dolomite-type structure (Δr_{Ca-Mg} = 0.28 Å), and for the $Ln^{3+}M^{3+}_3(BO_3)^{3-}_4$ with the Ln = Tb (r_{Tb}^{VI} = 0.92 Å) in combination with the M^{3+} = Sc (r_{Sc}^{VI} = 0.745 Å) (Δr_{Tb-Sc} = ~0.175 Å), forming the calcite-type structure. Indeed, a polycrystalline sample with the initial composition $Yb^{3+}Ga^{3+}_3(BO_3)^{3-}_4$, which was sintered between 575 and 1050 °C [34], and a single crystal with the initial composition $Tb^{3+}Sc^{3+}_3(BO_3)^{3-}_4$, obtained by the Czochralski method [35,83], crystallize with a decrease in symmetry, but with the same unit cell parameters, forming superstructures to dolomite with the space group R3 (a = 4.726(3), c = 15.43(2) Å) and calcite with the space group $R\bar{3}$ (a = 4.773(5), c = 15.48(1) Å), respectively. The XRD analysis of $Tb^{3+}Sc^{3+}_3(BO_3)^{3-}_4$ single crystal allowed to reveal additional diffraction reflections $h\bar{h}0l$ with the $h + l$ = $2n$, which are absent for the space group $R\bar{3}c$ and possible for the space groups $R\bar{3}$ and R3. For the crystals obtained, a non-synchronous second harmonic generation was not observed, which indicates the space groups $R\bar{3}$.

Single crystals with the initial composition $LnSc_3(BO_3)_4$ with the Ln = La [74,75], Ce [81], Pr [80,88], Nd [77,80,88], Sm [80,88], Eu [88], obtained by the flux method (the $NdSc_3(BO_3)_4$ were grown by the Czochralski method [77]), and also with the Ln = Gd [36,37] and Y [36,37], grown by the TSSG method, have a modification with the huntite structure (space group R32) (Figures 1 and 2). Fedorova et al. [78] as well as Li et al. [25] and Ye et al. [31] could not obtain a stable modification of $LaSc_3(BO_3)_4$ with the space group R32 by solid state reaction ($LaSc_3(BO_3)_4$ samples with the space group $C2/c$ have been obtained) and by a high-temperature solution method, respectively, suggesting that this phase seems to be metastable or stable in a narrow temperature range. It should be noted that the space group $R32$ for the $LnSc_3(BO_3)_4$ with the Ln = La [74,75], Ce [81], Pr [88], Nd [77,88], Sm [88], Eu [88] is determined by the single crystal XRD study, whereas for the Ln = Gd, (Y), the space group $R32$ is stated in [36,37] without any experimental confirmation, noting only that single crystals grown by the TSSG method exhibited well developed facets having the form of rhombohedral prisms characteristic of the space group $R32$.

For the crystals grown by the Czochralski method from the initial charges with the compositions $Pr_{1.1}Sc_{2.9}(BO_3)_4$ (PSB–1.1) and $Pr_{1.25}Sc_{2.75}(BO_3)_4$ (PSB–1.25) (Δr_{Pr-Sc} = 0.245 Å), $NdSc_3(BO_3)_4$ (NSB–1.0) and $Nd_{1.25}Sc_{2.75}(BO_3)_4$ (NSB–1.25) (Δr_{Nd-Sc} = 0.235 Å), a symmetry decrease from the space group $R32$ to $P321$ or $P3$ (Figures 1 and 2) was found by the XRD analysis (it should be noted that the value Δr_{Nd-Sc} (Å) is less than the critical value Δr_{Ln-M} = ~0.25 Å, at which the derivatives of the huntite structure are formed; Δr_{Pr-Sc} (Å) is actually at the stability limit). The extinction laws for an overwhelming number of diffraction reflections witness a crystallization of these compounds in the space group $R32$. However, 60% of the additional reflections with the $I \geq 3\sigma(I)$ are described within the framework of the superstructure having the huntite unit cell parameters, but with the space group $P321$ or $P3$ (the structures were solved in the space group $P321$) with the $h + k = 3n$, $l = 2n + 1$ for hkl [82]. In crystal structures with the space group $P321$ (Figure 4) compared with those with the space group $R32$ (Figure 3), the Ln and Sc crystallographic sites (Figure 3a–f), are split into two (Figure 4a–f), $Ln1$ and $Ln2$ (Figure 4g), sites with a distorted trigonal-prismatic oxygen environment and two, Sc1 and Sc2 (Figure 4h), sites with a distorted octahedral oxygen environment, respectively.

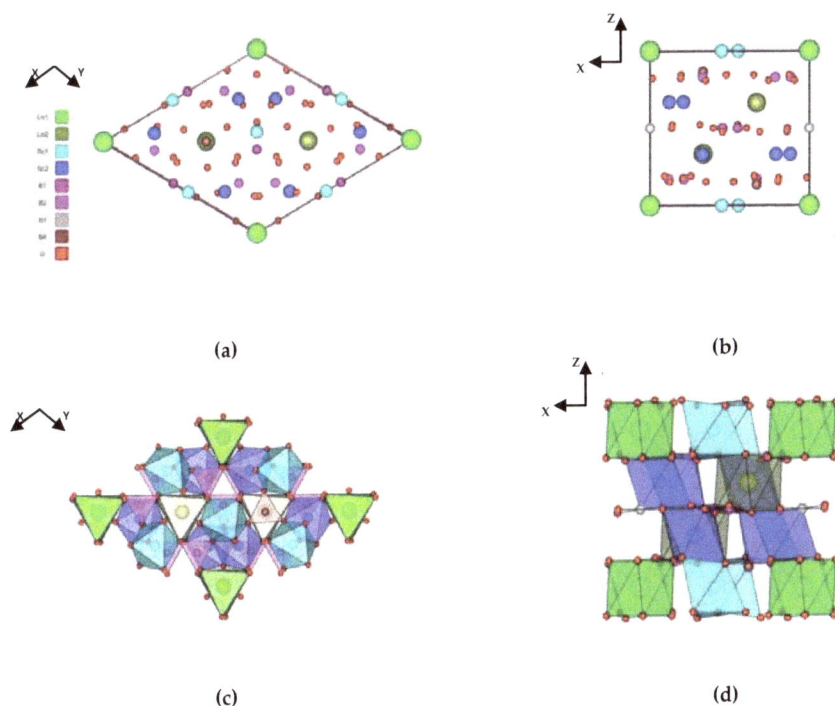

(a)

(b)

(c)

(d)

Figure 4. *Cont.*

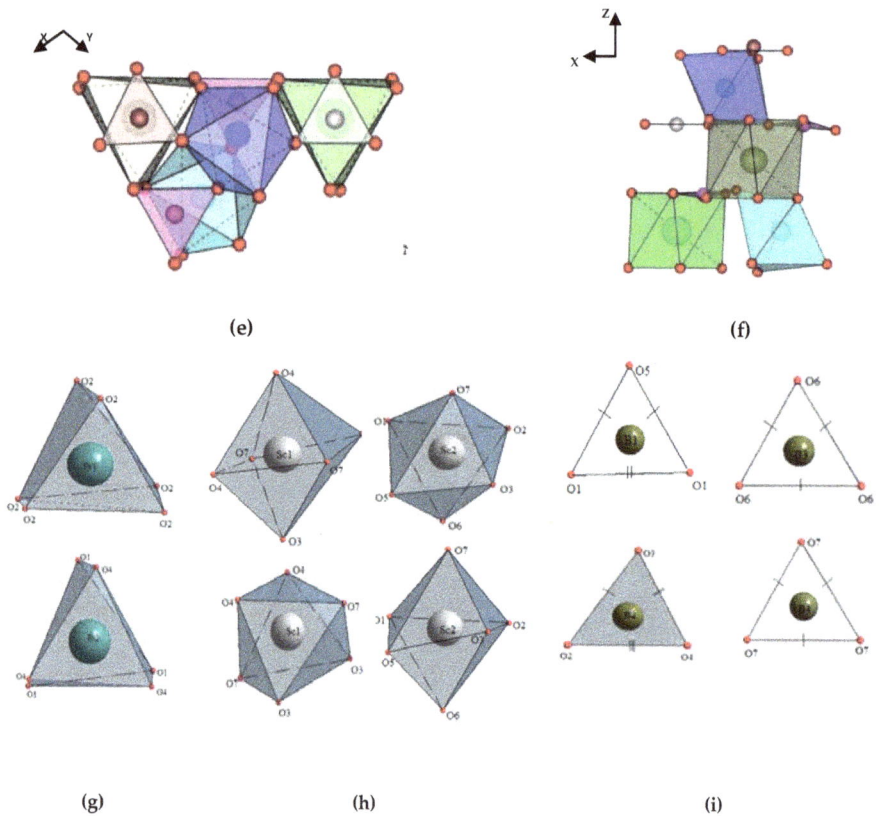

Figure 4. The unit cell of the $PrSc_3(BO_3)_4$ (PSB–1.1 and PSB–1.25) structure (space group $P321$) projected onto the (**a**) XY and (**b**) XZ planes; Combination of the coordination polyhedra projected onto the (**c**) XY and (**d**) XZ planes; Combination of selected coordination polyhedra projected onto the (**e**) XY and (**f**) XZ planes; Coordination polyhedra for the (**g**) Pr1 and Pr2, (**h**) Sc1 and Sc2, (**i**) B1–B4.

In addition, the number of B crystallographic sites is also increased in the structure with the space group $P321$ (Figure 4i) compared with the $R32$ one (Figure 3i). A comparison of the XZ projections of the structures with the space groups $R32$ (Figure 3b) and $P321$ (Figure 4b) indicates an alternation of layers of atoms La, Sc–B, O–La, Sc (Figure 3b) and La1, Sc1–B, O–La2, Sc2–B, O (Figure 4b) and corresponding polyhedra (Figure 3d,f and Figure 4d,f) along the Z axis.

The refined crystal compositions can be written as $[(Pr_{0.419}Sc_{0.081(4)})(1)]Pr_{0.5}(2)Sc_3(BO_3)_4$ $((Pr_{0.919}Sc_{0.081(4)})Sc_3(BO_3)_4)$ (PSB–1.1) and $[(Pr_{0.424}Sc_{0.076(4)})(1)]Pr_{0.5}(2)Sc_3(BO_3)_4$ $((Pr_{0.924}Sc_{0.076(4)})Sc_3(BO_3)_4)$ (PSB–1.25) [82], from which it follows that a symmetry decrease is caused by the distribution of (Pr, Sc) "atoms" and Pr atoms over two trigonal-prismatic sites (partial positional ordering) (Figure 4). Results of the XRD analysis with a refinement of positional and atom displacement parameters of single crystals with the initial composition $PrSc_3(BO_3)_4$, obtained by the flux method, are given in [88]. On the basis of systematic absences *hkil*: $-h + k + l \neq 3n$ and a successful refinement of the data for crystal, the space group was determined to be $R32$; the real composition of the crystal (i.e., the refinement of the site occupancies) was not determined [88].

Crystals with the initial compositions $NdSc_3(BO_3)_4$ (NSB–1.0) and $Nd_{1.25}Sc_{2.75}(BO_3)_4$ (NSB–1.25) with the space group $P321$ (in Figures 1 and 2, it is given as $P321^*$) have the refined compositions $NdSc_3(BO_3)_4$ and $(Nd_{0.500(1)}[Nd_{0.410}Sc_{0.090(20)}(2)]Sc_3(BO_3)_4$ $((Nd_{0.910}Sc_{0.090(20)})Sc_3(BO_3)_4)$, respectively.

These structures differ from each other (Figures 5 and 6) by an additional presence of the Sc ions in the trigonal-pyramidal Nd2 sites in the NSB–1.25 structure. It should be noted that the NSB structure is represented by right (NSB–1.0) (Figure 5a–f) and left (NSB–1.25) (Figure 6a–f) forms.

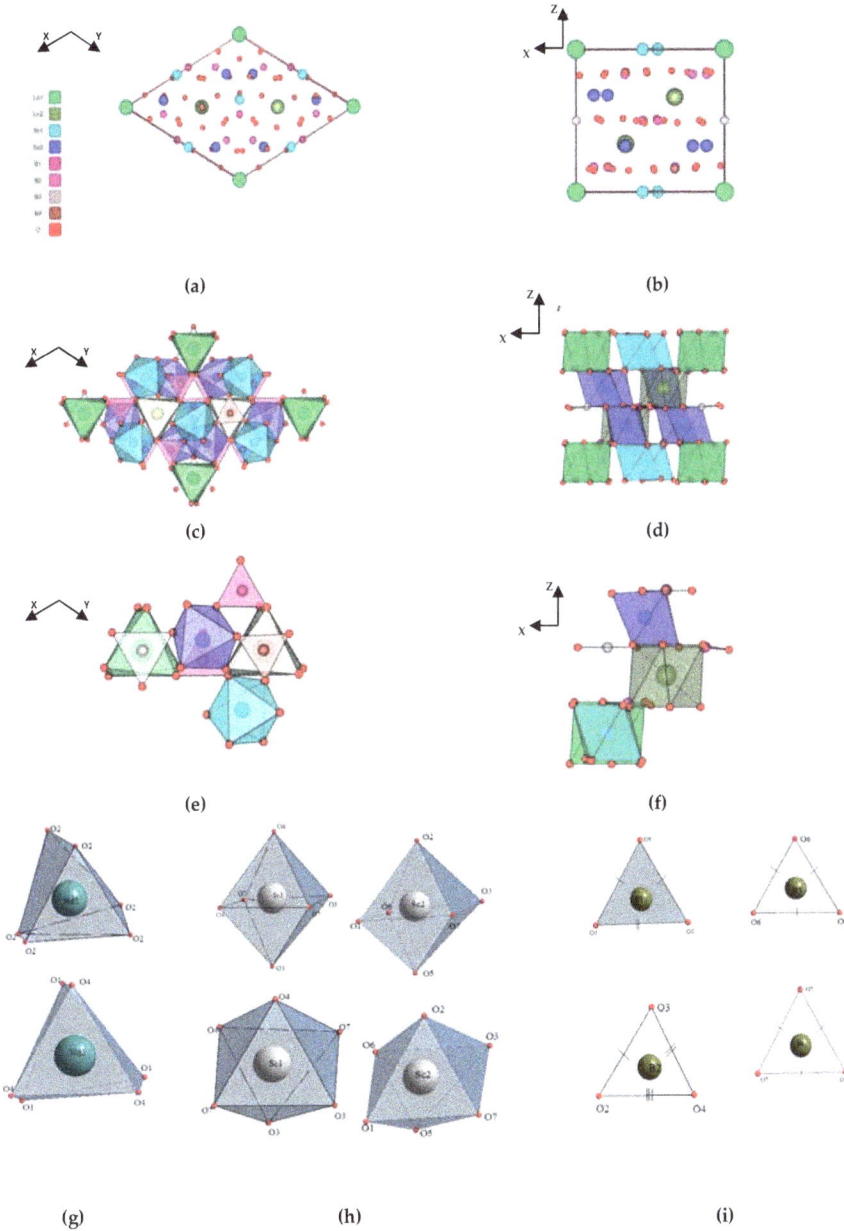

Figure 5. The unit cell of the $NdSc_3(BO_3)_4$ (NSB–1.0) structure (space group $P321$) projected onto the (a) XY and (b) XZ planes; Combination of the coordination polyhedra projected onto the (c) XY and (d) XZ planes; Combination of selected coordination polyhedra projected onto the (e) XY and (f) XZ planes; Coordination polyhedra for the (g) Nd1 and Nd2, (h) Sc1 and Sc2, (i) B1–B4.

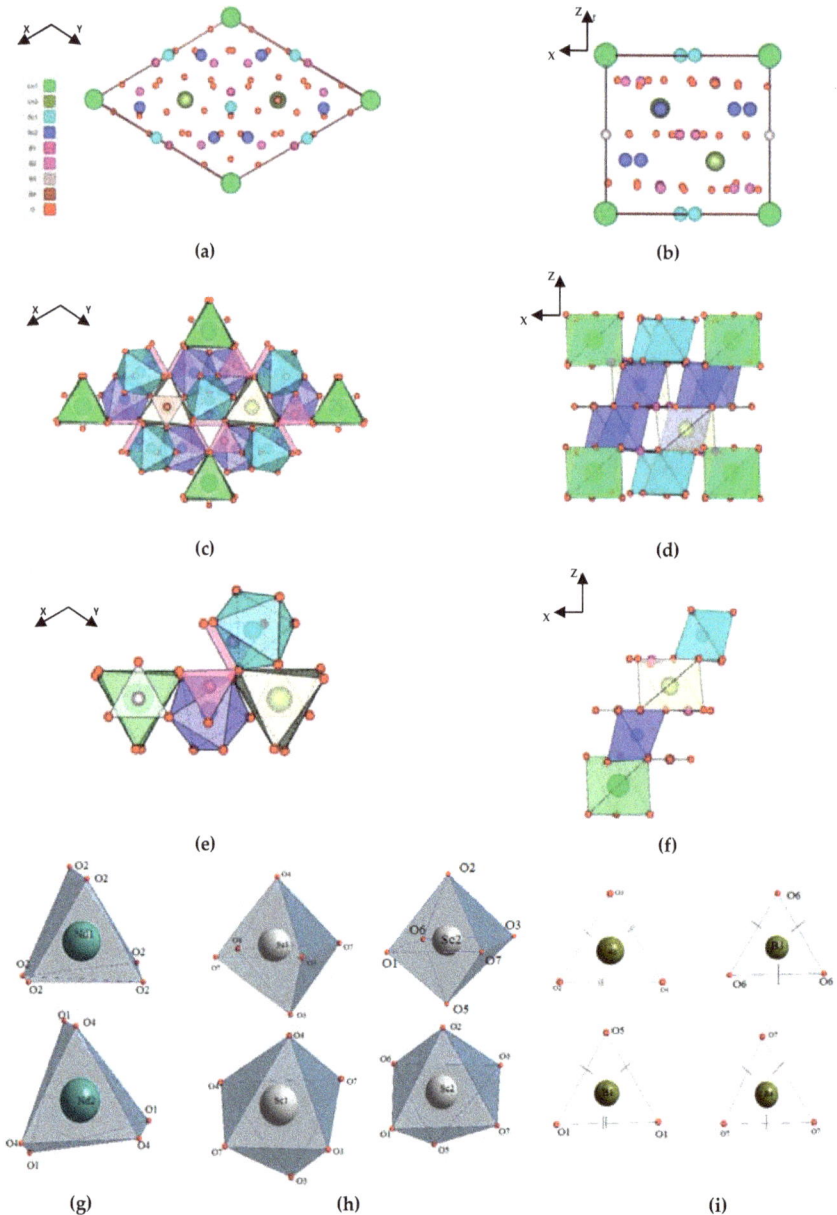

Figure 6. The unit cell of the NdSc$_3$(BO$_3$)$_4$ (NSB–1.25) structure (space group *P*321) projected onto the (**a**) XY and (**b**) XZ planes; Combination of the coordination polyhedra projected onto the (**c**) XY and (**d**) XZ planes; Combination of selected coordination polyhedra projected onto the (**e**) XY and (**f**) XZ planes; Coordination polyhedra for the (**g**) Nd1 and Nd2, (**h**) Sc1 and Sc2, (**i**) B1–B4.

The NdO$_6$ coordination polyhedra in the NSB–1.0 structure (Figure 5g) are more distorted than those in the NSB–1.25 one (Figure 6g), and the ScO$_6$ polyhedra in the NSB–1.0 structure (Figure 5h) are more regular (especially the Sc1O$_6$ polyhedron) than those in the NSB–1.25 one (Figure 6h). It confirms, firstly, a difference in their real compositions and, secondly, a 'stress removal' in the NSB–1.25 structure due to the presence of Sc atoms in trigonal-prismatic polyhedra.

Compared to the PSB structures (Figure 4), in the NSB–1.0 structure (Figure 5), the ScO_6 coordination polyhedron (Figures 4h and 5h) and one B polyhedron (Figures 4i and 5i) are less distorted, as well as the φ rotation angle in the $Ln2O_6$ coordination polyhedron is smaller (Figures 4g and 5g), which is caused by the presence of Sc atoms in the Pr site in the PSB structures.

As a result of the analysis of Figures 5 and 6, it can be concluded that the crystals grown by the Czochralski method from the charges with compositions $NdSc_3(BO_3)_4$ (NSB–1.0) and $Nd_{1.25}Sc_{2.75}(BO_3)_4$ (NSB–1.25) are characterized by the different structure disordering compared to the $Pr_{1.1}Sc_{2.9}(BO_3)_4$ (PSB–1.1) and $Pr_{1.25}Sc_{2.75}(BO_3)_4$ (PSB–1.25); NSB and PSB are isotypic structures (space group *P*321) (Figures 4–6). The differences are related to a distribution of the *Ln* and Sc ions over the trigonal-prismatic sites, resulting in another type of structure, namely, unequal $Nd1O_6$ and $Nd2O_6$ trigonal prisms in the NSB compared to the PSB structures and a different character of the changes in the interatomic distances in these polyhedra, primarily, the Nd(Pr)2–O4 и Nd(Pr)2–O1 ones [82]. The differences in the PSB–1.1, PSB–1.25, NSB–1.0, NSB–1.25 structures can be traced, for example, on their XZ projections (Figures 4b, 5b and 6b), paying attention to the B and O atoms.

Depending on the composition and growth conditions, the huntite-family compounds have monoclinic modifications (Figures 1 and 2) with the centrosymmetric space group *C*2/*c* (Figure 7) and non-centrosymmetric space groups *Cc* (the extinction laws are the same as those for the space group *C*2/*c*) (Figure 8) and *C*2 (Figure 9) (the extinction laws are the same as those for the space groups *C*2/*m*, *Cm*).

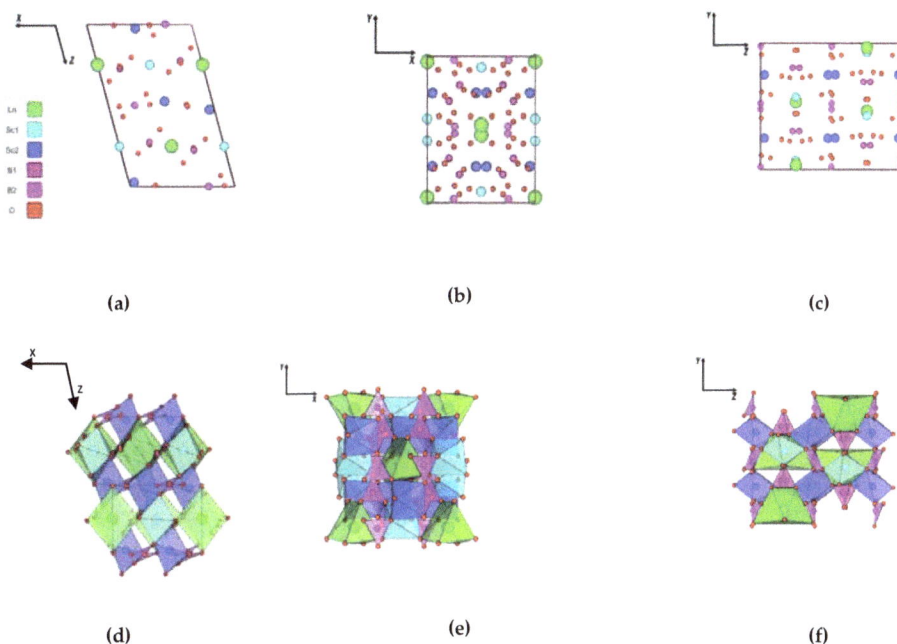

(a) (b) (c)

(d) (e) (f)

Figure 7. *Cont.*

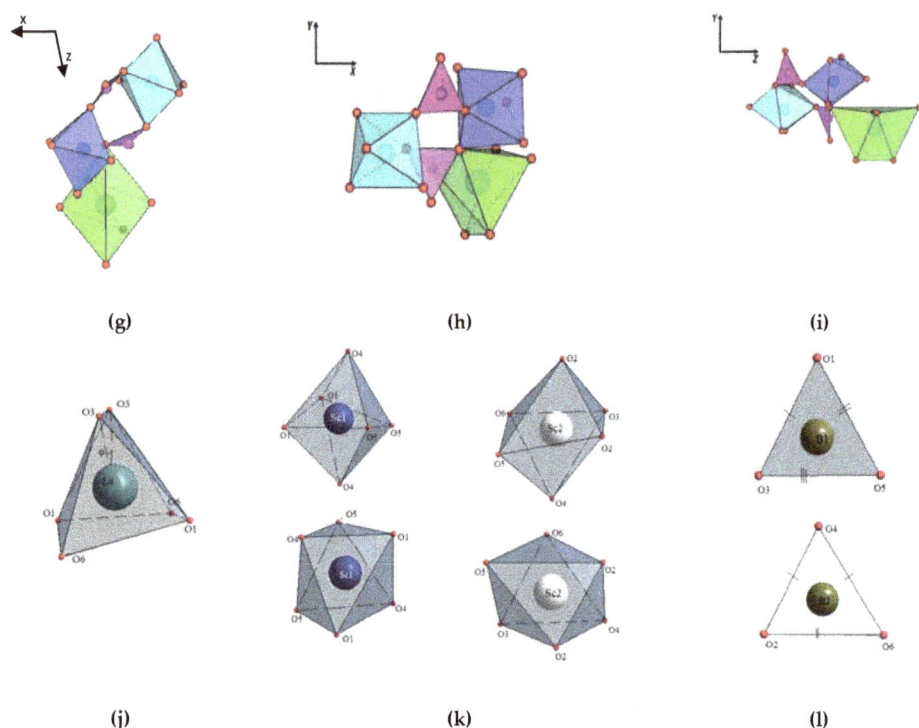

Figure 7. The unit cell of the $LnM_3(BO_3)_4$ structure (space group $C2/c$) projected onto the (**a**) XZ, (**b**) XY, (**c**) YZ planes; Combination of the coordination polyhedra projected onto the (**d**) XZ, (**e**) XY, (**f**) YZ planes; Combination of selected coordination polyhedra projected onto the (**g**) XZ, (**h**) XY, (**i**) YZ planes; Coordination polyhedra for the (**j**) *Ln*, (**k**) *M1* and *M2*, (**l**) *B1* and *B2*.

The space group $C2/c$ is known for the $LnAl_3(BO_3)_4$ with the Ln = Pr, Nd, Sm, Eu, Tb, Ho (the structure is refined, for example, in [97]); $LnFe_3(BO_3)_4$ with the Ln = Nd; $LnCr_3(BO_3)_4$ with the Ln = La, Pr, Nd, Sm, Tb, Dy, Ho; $LnSc_3(BO_3)_4$ with the Ln = La, Ce, Pr (Table 1). The centrosymmetry was determined by spectroscopic studies using a factor-group analysis of vibrations (for the $LnAl_3(BO_3)_4$ with the Ln = Pr [42], Nd [42,45]; $LnFe_3(BO_3)_4$ with the Ln = Nd [45]; $LnCr_3(BO_3)_4$ with the Ln = La (Figure 10), Pr, Tb, Dy, Ho [70,71], Nd [26,45,70,71], Sm [70,72]). The $LnSc_3(BO_3)_4$ crystals with the Ln = La, Ce, obtained by the Czochralski method, crystallize in the centrosymmetric space group $C2/c$), according to a study of their nonlinear optical properties [35]. The XRD investigation of the $CeSc_3(BO_3)_4$ microcrystal with the space group $C2/c$ indicates the similarity of charge and as-grown crystal compositions [82].

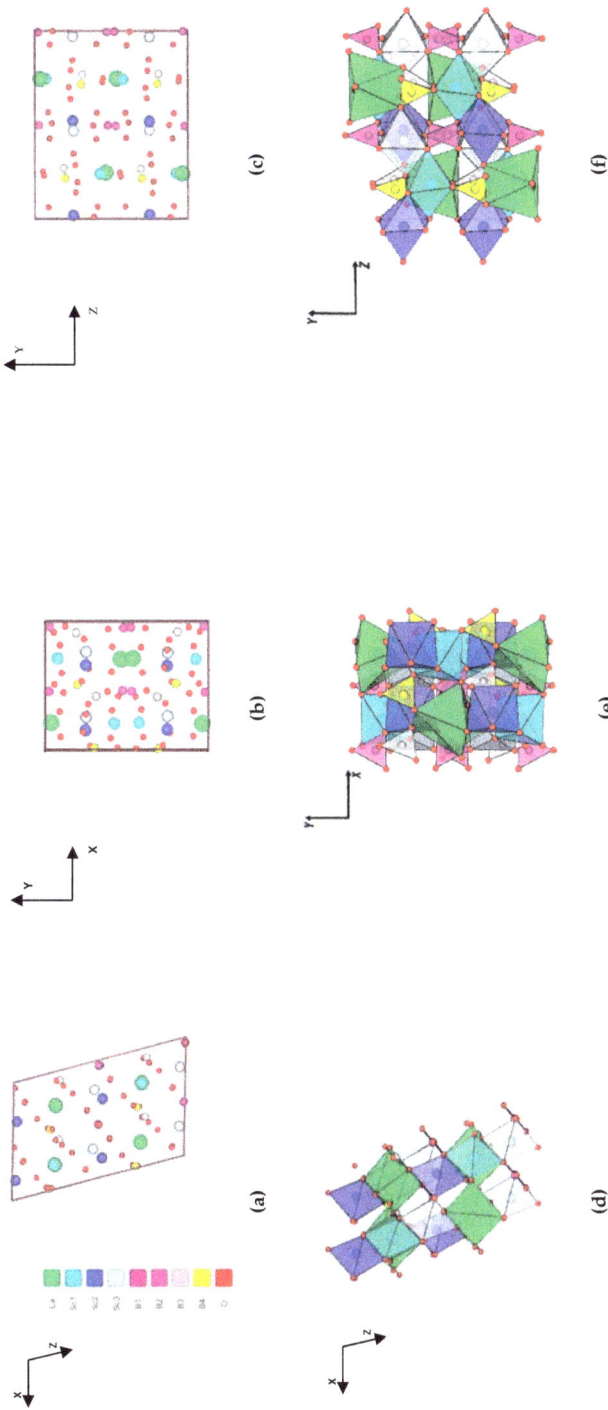

Figure 8. *Cont.*

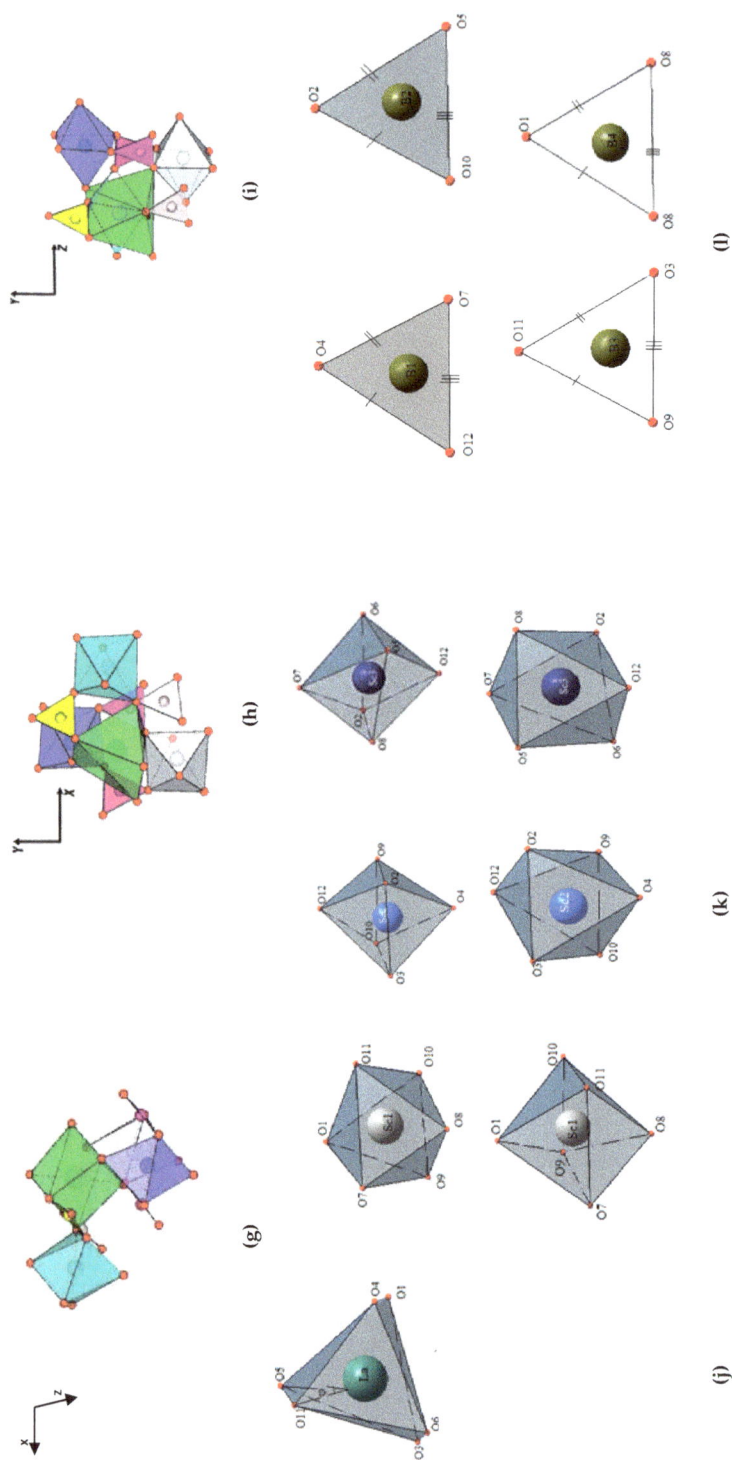

Figure 8. The unit cell of the *LnM*$_3$(BO$_3$)$_4$ structure (space group *Cc*) projected onto the (**a**) *XZ*, (**b**) *XY*, (**c**) *YZ* planes; Combination of the coordination polyhedra projected onto the (**d**) *XZ*, (**e**) *XY*, (**f**) *YZ* planes; Combination of selected coordination polyhedra projected onto the (**g**) *XZ*, (**h**) *XY*, (**i**) *YZ* planes; Coordination polyhedra for the (**j**) *Ln*, (**k**) *M1–M3*, (**l**) B1–B4.

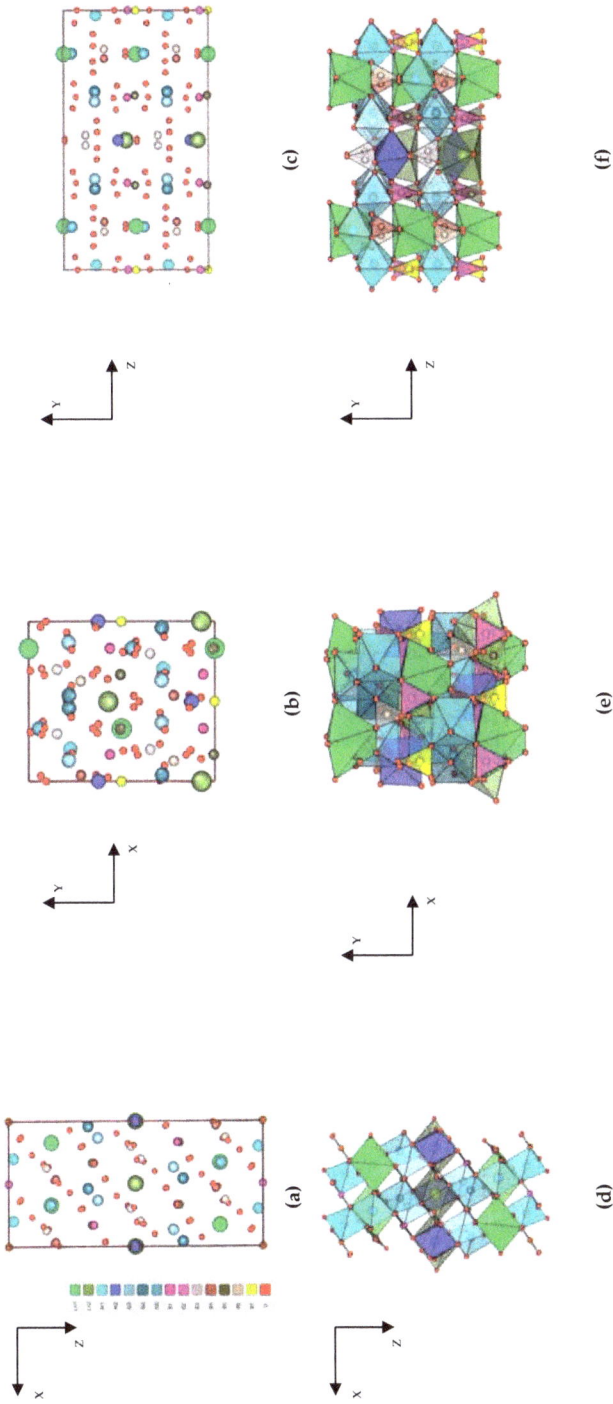

Figure 9. *Cont.*

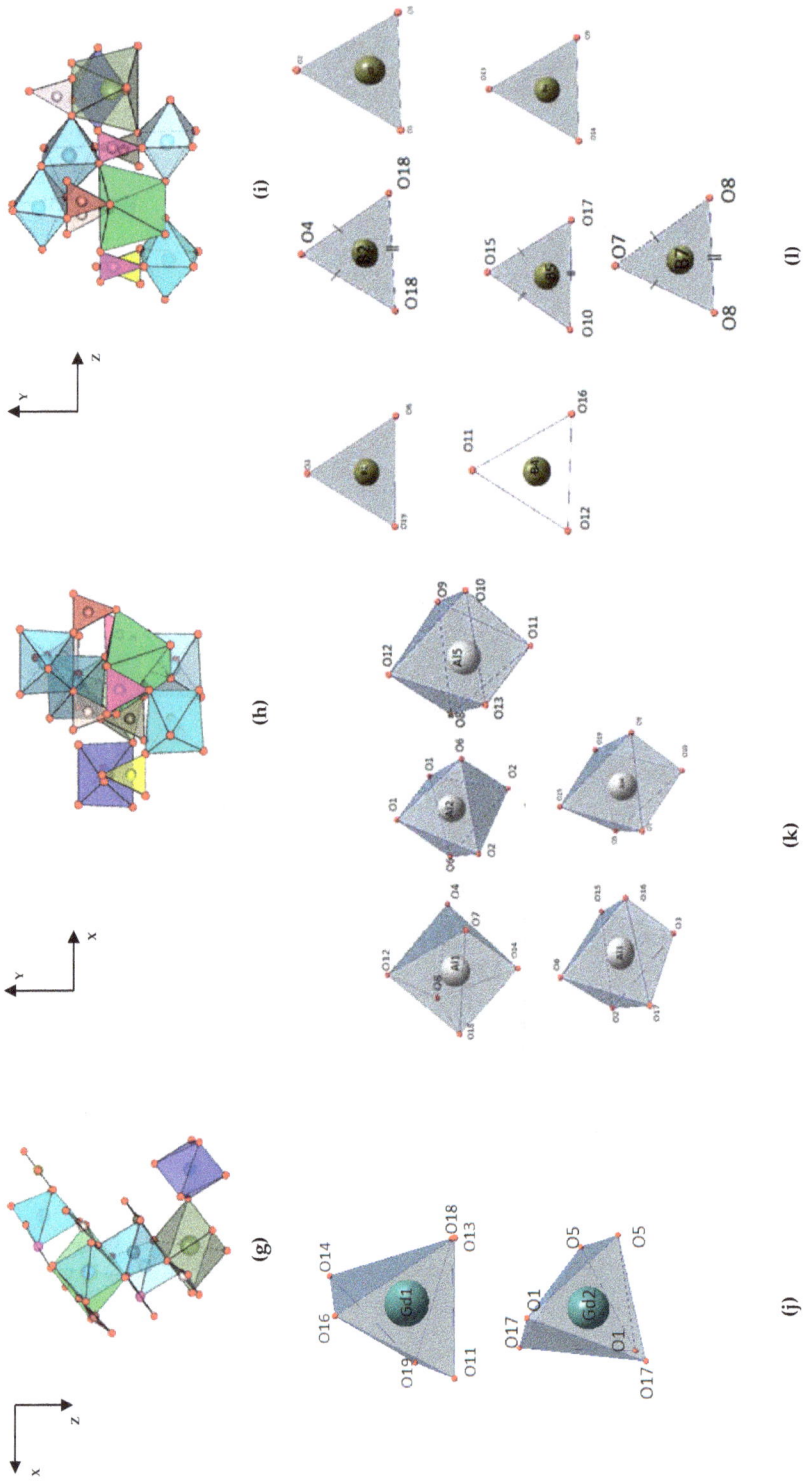

Figure 9. The unit cell of the *LnM*$_3$(BO$_3$)$_4$ structure (space group C2) projected onto the (**a**) XZ, (**b**) XY, (**c**) YZ planes; Combination of the coordination polyhedra projected onto the (**d**) XZ, (**e**) XY, (**f**) YZ planes; Combination of selected coordination polyhedra projected onto the (**g**) XZ, (**h**) XY, (**i**) YZ planes; Coordination polyhedra for the (**j**) *Ln1* and *Ln2*, (**k**) *M1–M5*, (**l**) B1–B7.

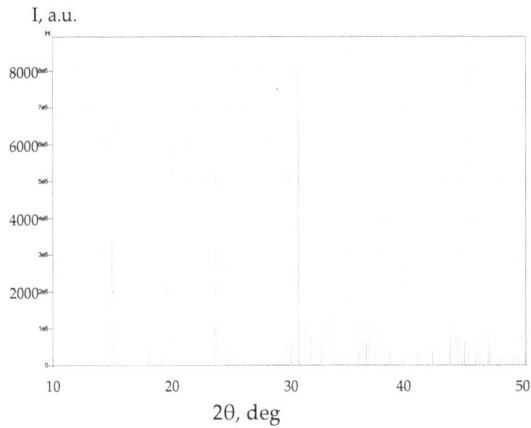

I, a.u.

(a)

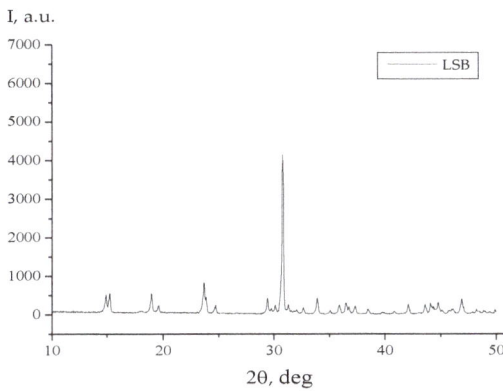

I, a.u.

(b)

Figure 10. (**a**) Theoretical and (**b**) experimental diffraction patterns for the α - LaSc$_3$(BO$_3$)$_4$ structure (space group *C2/c*).

In the crystal structure with the space group *C2/c* (Figure 7) the *Ln* atom is located at the center of a distorted trigonal prism (CN *Ln* = 2 + 2 + 2) (Figure 7j); the upper triangular face of prism is rotated with respect to the lower one by an angle φ = 8°. The Sc1 and Sc2 atoms are located in distorted octahedra (CN Sc1 = 2 + 2 + 2, CN Sc2 = 1 + 1 + 1 + 1 + 1 + 1) (Figure 7k), and the B1 and B2 atoms are at the centers of scalene triangles (CN B1 = 1 + 1 + 1; CN B2 = 1 + 1 + 1) (Figure 7l). A comparison of crystal structures with the space groups *R*32 (Figure 3) and *C2/c* (Figure 7) shows their similarity (the same set of polyhedra; however, in the structure with the space group *C2/c*, there are two crystallochemically-different Sc atoms rather that one as in the structure with the space group *R*32) (Figure 3g,h,I and Figure 7j,k,l), on the one hand, and differences (the polyhedra in the structure with the space group *C2/c* are more distorted), on the other hand, i.e., the structure with the space group *C2/c* is more disordered compared to the *R*32. A similar alternation of layers of atoms ... Ln, Sc–B, O–Ln, Sc and Ln, Sc2–B, O–Sc1–B, O–Ln, Sc1, Sc2–B, O in the structures with the space groups

*R*32 (XZ projection; Figure 3b) and *C*2/*c* (XZ projection; Figure 7a) and the corresponding polyhedra (Figure 3d,f and Figure 7d,g) should be mentioned.

The low temperature phase γ-LaSc$_3$(BO$_3$)$_4$ crystallizes in the space group *Cc* (*a* = 7.740(3), *b* = 9.864(2), *c* = 12.066(5) Å, β = 105.48(5)°) (Figure 8) [79], but there is no information on how its noncentrosymmetry was determined. The γ-LaSc$_3$(BO$_3$)$_4$ single crystal was grown by the TSSG method from a flux and its structure was solved by direct method and refined by full-matrix least squares without any refinement of real crystal composition [79]. In the structure with the space group *Cc* (Figure 8), which is positionally and structurally disordered compared with the α- LaSc$_3$(BO$_3$)$_4$ modification with the space group *C*2/*c* (Figure 7) the Sc1, Sc2, and Sc2 atoms are located in distorted octahedra (CN Sc = 1 + 1 + 1 + 1 + 1 + 1) (Figure 8k), and the B1, B2, B3, and B4 atoms are at the centers of scalene triangles (CN B = 1 + 1 + 1) (Figure 8l). According to Wang et al. [79], in the crystal structure of γ- LaSc$_3$(BO$_3$)$_4$ (space group *Cc*), the La atoms are at the centers of distorted octahedra with the CN La = 1 + 1 + 1 + 1 + 1 + 1. However, according to the rotation angle φ = 14° between the upper and lower faces (Figure 8j), which is larger than that in the structure with the space group *C*2/*c* (Figure 7j), the La polyhedron is a trigonal prism, since it is nominally considered that a polyhedron with the CN = 6 is a distorted trigonal prism or a distorted octahedron in the angle ranges 0° < φ <<~ 30° or ~30° < φ < 60°, respectively. Actually, based on a structure with the space group *C*2/*c*, the Sc2 crystallographic site (Figure 7a–i) is 'split' into two sites, Sc2 and Sc3 (Figure 8a–i), the B1 site (Figure 7l) is 'split' into the B1 and B2 sites (Figure 8l), and the B2 site (Figure 7l) is 'split' into the B3 and B4 sites (Figure 8l) with the formation of a structure with the space group *Cc*. Hence, a transition from the centrosymmetric structure with the space group *C*2/*c* to non-centrosymmetric structure with the space group *Cc* is obliged to 'split' sites (see, for example, Figures 7f and 8f). In the structures with the space groups *C*2/*c* and *Cc*, a similar alternation of layers of atoms and polyhedra with these atoms is observed on XZ projections: Ln, Sc2–B, O–Sc1–B, O–Ln, Sc1, Sc2–B, O (Figure 7a,d) and Ln, Sc1, Sc2–B, O–Ln, Sc2, Sc3–B, O–Ln, Sc2, Sc3–B, O–Ln, Sc1, Sc3–B, O (Figure 8a,d).

Fedorova et al. [78], based on a complete correspondence (the space groups *C*2/*c* and *Cc* belong to the same diffraction symmetry group with the same extinction laws) of the diffraction patterns of powdered α-LaSc$_3$(BO$_3$)$_4$ (space group *C*2/*c*) and γ- LaSc$_3$(BO$_3$)$_4$ (space group *Cc*), obtained by solid state synthesis and heat-treated at temperatures from 1000 to 1350 °C, concluded that the LaSc$_3$(BO$_3$)$_4$ crystallizes in the space group *C*2/*c*. The experiment described does not allow choosing one of two space groups correctly.

The unit cell parameters of single crystal grown from the charge having composition (Pr$_{1.1}$Sc$_{2.9}$)(BO$_3$)$_4$ (*a* = 7.7138(6), *b* = 9.8347(5), *c* = 12.032(2) Å, β = 105.38(7)°) [83] are similar to those for the LaSc$_3$(BO$_3$)$_4$ and CeSc$_3$(BO$_3$)$_4$ (*a* = 7.7297(3), *b* = 9.8556(3), *c* = 12.0532(5)Å, β = 105.405(3)°) with the space group *C*2/*c* [82]. The extinction laws for the overwhelming number of diffraction reflections correspond to the space group *C*2/*c* or *Cc* (*h* + *k* = 2n for *hkl*, *h* = 2n, *l* = 2n for *h0l*, *k* = 2n for 0*kl*, *h* + *k* for *hk*0, *h* = 2n for *h*00, *k* = 2n for 0*k*0, *l* = 2n for 00*l*). Non-synchronous second harmonic generation was observed for these crystals, which indicates a non-centrosymmetric space structure *Cc*. A small amount of additional reflections with the *I* ≥ 3σ(*I*), typical for the space groups *C*2/*m*, *C*2 or *Cm* (*h* + *k* = 2n for *hkl*, *h* = 2n for *h0l*, *k* = 2n for 0*kl*, *h* + *k* for *hk*0, *h* = 2n for *h*00, *k* = 2n for 0*k*0), most likely, space group *C*2 as a subgroup of *C*2/*c*, is observed.

The space group *Cc* was not found for the *Ln*M$_3$(BO$_3$)$_4$ with the *M* = Al, Fe, Cr, Ga. For the *Ln*Al$_3$(BO$_3$)$_4$ with the *Ln* = Pr, Nd, Eu, Gd, a modification with the space group *C*2 (*a* = 7.262(3), *b* = 9.315(3), *c* = 16.184(8)Å, β = 90.37° for the GdAl$_3$(BO$_3$)$_4$) [47] was revealed. In the crystal structure of *Ln*Al$_3$(BO$_3$)$_4$ with the space group *C*2 (Figure 9), five crystallochemically-different Al atoms are located at distorted octahedra (CN Al1, Al3, Al4, Al5 = 1 + 1 + 1 + 1 + 1 + 1, CN Al2 = 2 + 2 + 2) (Figure 9k) and seven crystallochemically-different B atoms have a trigonal environment, two of which are isosceles triangles and the rest ones are scalene (Figure 9l) [41]. The *Ln* (Gd) atoms occupy distorted trigonal prisms with the CN Ln1 = 1 + 1 + 1 + 1 + 1 + 1 and CN Ln2 = 2 + 2 + 2, and the rotation angle is φ ~ 20° (Figure 9j), i.e., significantly higher than that for other modifications of huntite-like structures.

This feature of the $LnAl_3(BO_3)_4$ crystal structure with the space group $C2$, which is the most disordered among all the huntite-like structures described (Figure 9a–i), does not exclude a tendency to reorganize a trigonal prism into an octahedron, accompanied by a decrease in the polyhedron volume and degree of its filling [98,99]. It may be the reason or one of the reasons for an absence of this structure for the $LnSc_3(BO_3)_4$ compounds. Nevertheless, a comparison of monoclinic compounds with the space groups $C2/c$ (Figure 7), Cc (Figure 8), and $C2$ (Figure 9) indicates their identical structural features: 'layers' of atoms and polyhedra (Figure 7a,d, Figure 8a,d, Figure 9a,d), a connection of the LnO_6 trigonal prisms and $ScO6$ octahedra through the BO_3 triangles (Figure 7g–i, Figure 8g–i, Figure 9g–i). The same structural features can also be observed for the compounds with the space groups $R32$ (Figure 3) and $P321$ (Figures 4–6).

Based on the experimental data available to date, it is possible to limit the region of stability of the phases with the general composition $LnM_3(BO_3)_4$, having the huntite-like structures, as a correlation between the type of Ln ion and the difference between the ionic radii of Ln and M ions (Figure 11).

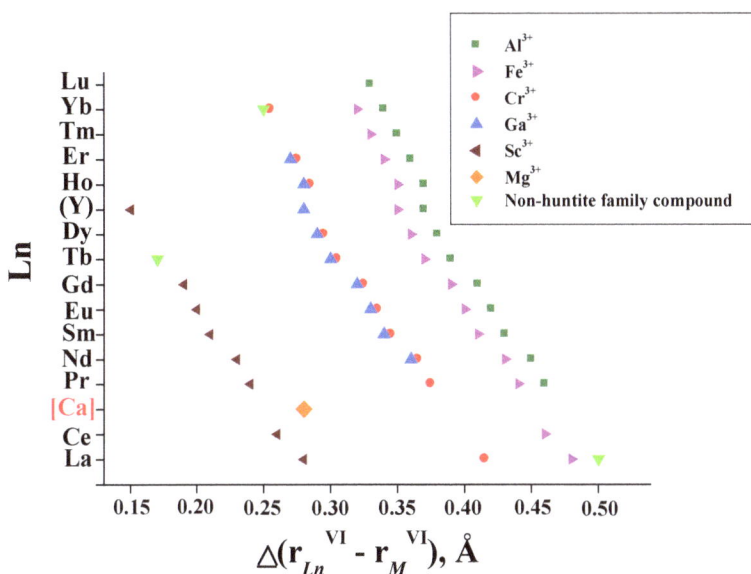

Figure 11. The stability region for the phases with the general composition $LnM_3(BO_3)_4$ with the M^{3+} = Al, Fe, Cr, Ga, Sc given as correlation between the type of Ln ion and the difference between the ionic radii of Ln and M ions.

In the coordinates given in Figure 11, the huntite $CaMg_3(CO_3)_4$ (space group $R32$) is located closer to the $LnSc_3(BO_3)_4$, which may explain a formation of superstructures with full or partial ordering of atoms over the crystallographic sites (positional ordering) accompanied by a symmetry decrease in both $CaCO_3$-$MgCO_3$ and Ln_2O_3-Sc_2O_3-B_2O_3 system. In addition, there are other structural features of rare-earth scandium orthoborates.

For polycrystalline and single-crystal samples with the general composition $LnSc_3(BO_3)_4$, for which the $r_{Ln}{}^{VI}$-$r_{Sc}{}^{VI}$ (Å) value is in the range of 0.285 (Ln = La) – 0.155 (Ln = Y) (i.e., with the minimum Δr_{Ln-M} (Å) values among all the huntite-like structures) (Figure 11), a probability of obtaining compounds with the stoichiometric composition $LnSc_3(BO_3)_4$ decreases, starting with the Ln = Nd (Δr_{Nd-Sc} = 0.235 Å). A distinctive feature of samples with the Ln = Sm (Δr_{Sm-Sc} = 0.215 Å), Eu (Δr_{Eu-Sc} = 0.215 Å), and Gd (Δr_{Gd-Sc} = 0.195 Å) should be a formation of internal solid solutions in the form $(Ln,Sc)Sc_3(BO_3)_4$ (matrix ions are redistributed over different crystallographic sites resulting in a formation of antisite defects). In samples with the Ln from Tb (Δr_{Tb-Sc} = 0.175 Å) to Lu (Δr_{Lu-Sc} = 0.110 Å), internal solid solutions in the form $(Ln,Sc)(Sc,Ln)_3(BO_3)_4$ will probably be formed. A formation of internal solid solutions can lead to a decrease in symmetry due to a positional disordering of the Ln and Sc cations, which may ultimately results in the compounds going beyond the limits of the stability region of the huntite-family phases (Figure 11).

The data on the symmetry of compounds with the general composition $LnSc_3(BO_3)_4$ are contradictory, in particular, for the modifications with the space group *R32*. A difficulty of revealing the space group *P321*, which was found for the PSB and NSB as a result of a comparison the experimental diffraction patterns with those given in databases and by the refinement of crystal structures of polycrystalline samples or powdered single crystals by the full-profile method (it is usually used in practice), is associated with the almost complete similarity of theoretical diffraction patterns for the space groups *R32* and *R321* (Figure 12a,b). The differences relate only to the intensities of individual reflections, shown in Figure 12a,b in a circle.

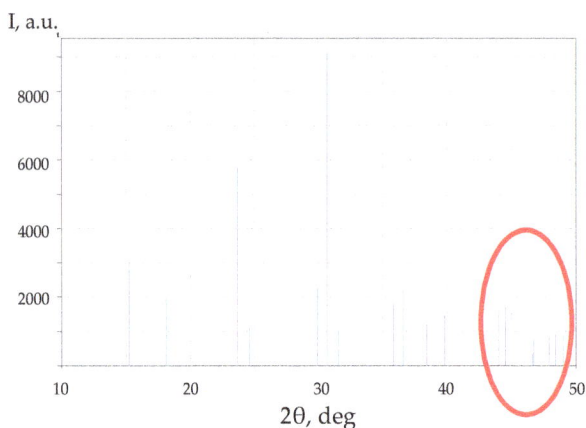

(a)

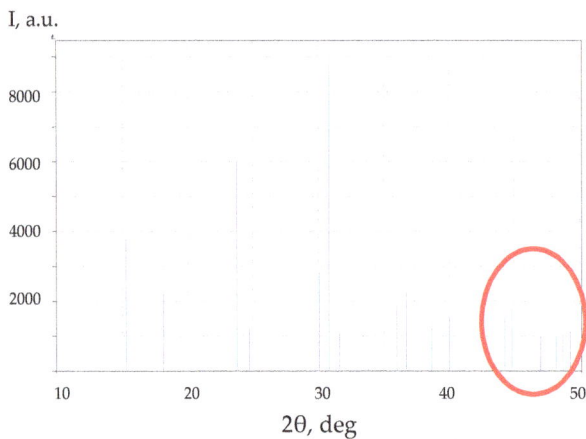

(b)

Figure 12. *Cont.*

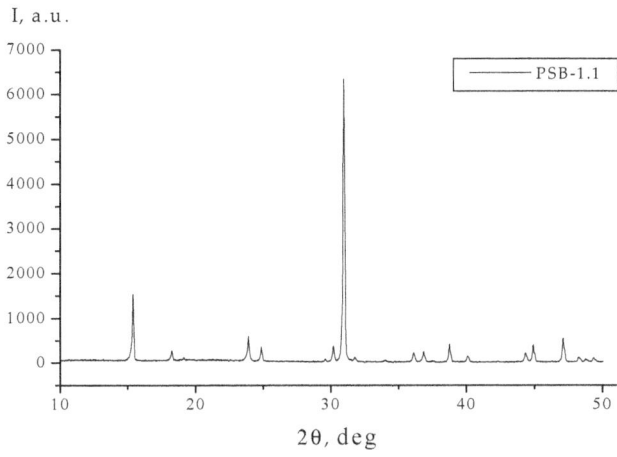

(c)

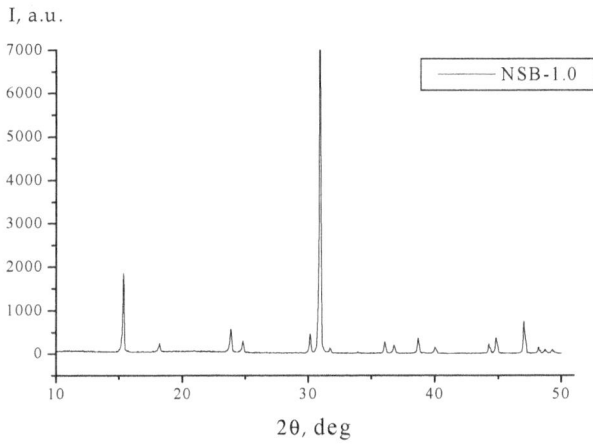

(d)

Figure 12. *Cont.*

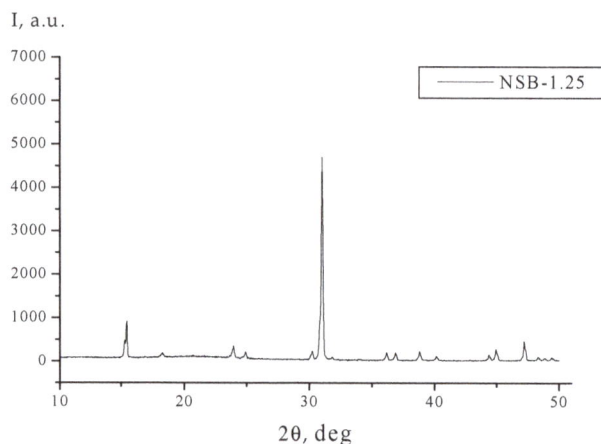

(e)

Figure 12. Theoretical diffraction patterns for the (**a**) huntite structure (space group *R*32), (**b**) superstructure of the huntite structure (space group *P*321), (**c**) structure with the space group $P3_121$; Experimental diffraction patterns of powdered single crystals (space group *P*321): (**c**) $Pr_{1.1}Sc_{2.9}(BO_3)_4$ (PSB–1.1), (**d**) $NdSc_3(BO_3)_4$ (NSB–1.0), (**e**) $Nd_{1.25}Sc_{2.75}(BO_3)_4$ (NSB–1.25).

Figure 12d–f show the experimental diffraction patterns of powdered PSB–1.1, NSB–1.0, and NSB–1.25 crystals with the space group *P*321, from which their similarity (with a certain redistribution of the intensities of a series of reflections) and an analogy with the theoretical diffraction patterns is also seen. In order to establish the space group, *R*32 and *P*321, in which the rare-earth scandium orthoborates crystallize, it is necessary to perform, for example, a single-crystal XRD experiment with an analysis of intensities of reflections, as was performed for the PSB and NSB crystals [82].

3.2. Solid Solutions of Rare-Earth Scandium Borates

The symmetry of solid solutions in the $LnSc_3(BO_3)_4 – Ln'Sc_3(BO_3)_4$, $LnSc_3(BO_3)_4 – Ln'Sc_3(BO_3)_4 – ScSc_3(BO_3)_4$ (space group $R\bar{3}c$), $LnSc_3(BO_3)_4 – Ln'Sc_3(BO_3)_4 – Ln''Sc_3(BO_3)_4$ systems is determined by the space group and the ratio of the system components. In the $LaSc_3(BO_3)_4$-$CeSc_3(BO_3)_4$ system, as well as in the $LaSc_3(BO_3)_4 – PrSc_3(BO_3)_4$ and $CeSc_3(BO_3)_4 – PrSc_3(BO_3)_4$ systems, provided that the $PrSc_3(BO_3)_4$ crystallizes in the space group *C*2/*c*, a formation of a continuous series of solid solutions with the space group *C*2/*c* should be expected. Limited solid solutions, which can be based on the compounds that actually exist, are formed in systems where the final members of solid solution belong to different symmetries (monoclinic and trigonal) (Figure 11). This obvious crystallochemical conclusion, based on the theory of isomorphic mixing, is confirmed experimentally (Table 2).

Table 2. Summary structural data on solid solutions of rare-earth scandium borates.

Initial Composition	Synthesis Method [1]	Space Group	Refined Composition (Method) [2]	Unit Cell Parameters, a, c, Å	Reference
LaSc$_3$(BO$_3$)$_4$–NdSc$_3$(BO$_3$)$_4$ System					
(La$_{1-x}$Nd$_x$)Sc$_3$(BO$_3$)$_4$ (x = 1.0)	SSR and Flux	R32			[25]
(La$_{1-x}$Nd$_x$)Sc$_3$(BO$_3$)$_4$ ($x \leq$ 0.5)	SSR and Flux	R32 or C2/c (depending on the crystallized temperature)			[25]
(La$_{1-x}$Nd$_x$)Sc$_3$(BO$_3$)$_4$ (x = 0.0–0.3)	Czochralski	C2/c			[35,87]
(La$_{1-x}$Nd$_x$)Sc$_3$(BO$_3$)$_4$ ($x \geq$ 0.5)	Czochralski	R32			[100]
LaSc$_3$(BO$_3$)$_4$–αGdSc$_3$(BO$_3$)$_4$» System					
La$_x$Gd$_{1-x}$Sc$_3$(BO$_3$)$_4$ (0.20 ≤ x ≤ 0.80)	SSR	R32			[101,102]
La$_{1-x}$Gd$_x$Sc$_3$(BO$_3$)$_4$ (0.3 ≤ x ≤ 0.7)	Flux	R32			[88]
La$_{1-x}$Gd$_x$Sc$_3$(BO$_3$)$_4$ ($x \geq$ 0.3)	SSR	R32			[102,104]
La$_{0.6}$Gd$_{0.4}$Sc$_3$(BO$_3$)$_4$	TSSG	R32	La$_{0.7}$Gd$_{0.22}$Sc$_{3.01}$(BO$_3$)$_4$		[101]
La$_{0.4}$Gd$_{0.6}$Sc$_3$(BO$_3$)$_4$	TSSG	R32	La$_{0.64}$Gd$_{0.35}$Sc$_{3.01}$(BO$_3$)$_4$ (ICP)		[101]
La$_{0.2}$Gd$_{0.8}$Sc$_3$(BO$_3$)$_4$	TSSG	R32	La$_{0.46}$Gd$_{0.56}$Sc$_{2.98}$(BO$_3$)$_4$ (ICP)		[101]
La$_x$Gd$_{1-x}$Sc$_3$(BO$_3$)$_4$	TSSG	R32	La$_{0.78}$Gd$_{0.22}$Sc$_3$(BO$_3$)$_4$ (ICP)	9.7933 7.9540	[101]
La$_{0.678}$Gd$_{0.572}$Sc$_{2.75}$(BO$_3$)$_4$	Czochralski	R32	La$_{0.64}$Gd$_{0.41}$Sc$_{2.95}$(BO$_3$)$_4$ (ICP)	9.794(4) 7.961(6)	[103]
La$_{0.75}$Gd$_{0.5}$Sc$_{2.75}$(BO$_3$)$_4$	Czochralski	R32		9.791 7.952	[104]
LaSc$_3$(BO$_3$)$_4$–αYSc$_3$(BO$_3$)$_4$» System					
Y$_x$La$_y$Sc$_z$(BO$_3$)$_4$ (0.29 < x < 0.67, 0.67 < y < 0.82, 2.64 < z < 3.00)	SSR	R32			[31,102,105]
La$_{0.77}$Y$_{0.28}$Sc$_{2.95}$(BO$_3$)$_4$, La$_{0.76}$Y$_{0.32}$Sc$_{2.92}$(BO$_3$)$_4$, La$_{0.80}$Y$_{0.38}$Sc$_{2.82}$(BO$_3$)$_4$, La$_{0.73}$Y$_{0.42}$Sc$_{2.85}$(BO$_3$)$_4$, La$_{0.75}$Y$_{0.47}$Sc$_{2.78}$(BO$_3$)$_4$	TSSG	R32			[31,102,105]
Y$_2$O$_3$:La$_2$O$_3$:Sc$_2$O$_3$:B$_2$O$_3$:Li$_2$O = 0.60.35:1.5:7:7.5	TSSG	R32	La$_{0.72}$Y$_{0.57}$Sc$_{2.71}$(BO$_3$)$_4$ (ICP)	9.774(1) 7.944(3)	[31,102,105–107]
Y$_x$La$_{1-x}$Sc$_3$(BO$_3$)$_4$	Flux	R32	La$_{0.75}$Y$_{0.25}$Sc$_3$(BO$_3$)$_4$ (XRPD)	9.805(3) 7.980(2)	[108]
La$_{0.72}$Y$_{0.57}$Sc$_{2.71}$(BO$_3$)$_4$	TSSG	R32	La$_{0.826}$Y$_{0.334}$Sc$_{2.84}$(BO$_3$)$_4$ (ICP)	9.8185 7.9893	[109]

Table 2. *Cont.*

Initial Composition	Synthesis Method [1]	Space Group	Refined Composition (Method) [2]	Unit Cell Parameters, a, c, Å	Reference
$LaSc_3(BO_3)_4$–α-$ErSc_3(BO_3)_4$ »*System*					
$LaEr_{0.006}Sc_{2.994}(BO_3)_{3.8}$, $LaEr_{0.015}Sc_{2.985}(BO_3)_{3.8}$	Czochralski	$C2/c$			[35]
$LaSc_3(BO_3)_4$–α-$YbSc_3(BO_3)_4$ »*System*					
$LaYb_{0.15}Sc_{2.85}(BO_3)_{3.8}$, $LaYb_{0.3}Sc_{2.7}(BO_3)_{3.8}$, $LaYb_{0.36}Sc_{2.64}(BO_3)_{3.8}$	Czochralski	$C2/c$			[35]
$LaSc_3(BO_3)_4$–α-$LuSc_3(BO_3)_4$ »*System*					
$La_xLu_ySc_z(BO_3)_4$ ($x+y+z=4$)	TSSG	$R32$	$(Lu_{0.05}La_{0.95})(Lu_{0.61}Sc_{2.39})(BO_3)_4 \equiv La_{0.95}Lu_{0.66}Sc_{2.39}(BO_3)_4$ (ICP)	9.87420(10) 8.0696(7)	[110]
$LaSc_3(BO_3)_4$–α-$BiSc_3(BO_3)_4$ »*System*					
$La_xBi_ySc_z(BO_3)_4$ ($0.27 < x < 0.52$, $0.67 < y < 0.82$, $2.74 < z < 2.95$)	SSR	$R32$			[111]
$La_2O_3 : Sc_2O_3 : Bi_2O_3 : B_2O_3 = 1 : 1.5 : 13 : 13$	Flux	$R32$	$La_{0.82}Bi_{0.27}Sc_{2.91}(BO_3)_4$ (ICP)	9.828(4) 7.989(7)	[111]
$La_2O_3 : Sc_2O_3 : Bi_2O_3 : B_2O_3 = 1 : 1.5 : 13 : 13$	Flux	$R32$	$La_{0.91}Bi_{0.21}Sc_{2.88}(BO_3)_4$ (EDS)	9.8370(14) 7.9860(14)	[112]
$CeSc_3(BO_3)_4$–$NdSc_3(BO_3)_4$ »*System*					
$Ce_{0.33}Nd_{0.64}Sc_{2.83}(BO_3)_4$	Czochralski	$P321$	$(Ce_{0.5}Nd_{0.5(1)})Sc_3(BO_3)_4$ (XRD; $P321$ [3])		[35]
$CeSc_3(BO_3)_4$–α-$GdSc_3(BO_3)_4$ »*System*					
$(Ce_{0.7}Gd_{0.3})Sc_3(BO_3)_4$	Czochralski	$P321$			present work
$(Ce_{0.8}Gd_{0.2})Sc_3(BO_3)_4$	Czochralski	$P321$	$(Ce_{0.485(3)}Gd_{0.009}Sc_{0.006})(Ce_{0.465(6)}Gd_{0.017}Sc_{0.018})Sc_3(BO_3)_4$ (XRD; $P321$ [3]) $(Ce_{0.807(2)}Gd_{0.091(4)}Sc_{0.106(4)})Sc_3(BO_3)_4$ (XRD; $R32$ [3])	9.7812(10) 7.9480(12)	present work
$Ce_{0.9}Gd_{0.35}Sc_{2.75}(BO_3)_4$	Czochralski	$P321$	$(Ce_{0.94(1)}Gd_{0.024(1)}Sc_{0.027(1)})Sc_3(BO_3)_4$ (XRD; $R32$ [3])	9.7776(42) 7.9436(21)	present work
$CeSc_3(BO_3)_4$–α-$YSc_3(BO_3)_4$ »*System*					
$Ce_{1.25}Y_{0.3}Sc_{2.45}(BO_3)_4$	Czochralski	$P321$	$(Ce_{0.78}Y_{0.22(2)})Sc_3(BO_3)_4$ (XRPD; $R32$ [3])		[35]
$Ce_{1.25}Y_{0.3}Sc_{2.45}(BO_3)_4$	Czochralski	$P321$	$(Ce_{0.95(1)}Y_{0.032(1)}Sc_{0.003(1)})Sc_3(BO_3)_4$ (XRD; $R32$ [3])	9.7553(25) 7.9680(12)	present work

Table 2. Cont.

Initial Composition	Synthesis Method [1]	Space Group	Refined Composition (Method) [2]	Unit Cell Parameters, a, c, Å	Reference
$CeSc_3(BO_3)_4$–$\alpha LuSc_3(BO_3)_4$ System					
$Ce_{1.25}Lu_{0.3}Sc_{2.45}(BO_3)_4$	Czochralski	P321	$Ce(Lu_{0.170})Sc_{2.83})(BO_3)_4$ (XRPD; R32 [3])		[35]
$Ce_{1.25}Lu_{0.3}Sc_{2.45}(BO_3)_4$	Czochralski	P321	$Ce(Sc_{2.910(30)}Lu_{0.090})(BO_3)_4$ (XRD; R32 [3])	9.8085(21) 7.9829(10)	present work
$PrSc_3(BO_3)_4$–$NdSc_3(BO_3)_4$ System					
$Pr_{0.99}Nd_{0.11}Sc_{2.9}(BO_3)_4$	Czochralski	P321			[35]
$PrSc_3(BO_3)_4$–$\alpha YSc_3(BO_3)_4$ System					
$Nd_2O_3 : Y_2O_3 : Sc_2O_3 : HBO_2 : Li_2(CO_3) : LiF = 0.25 : 0.25 : 0.8 : 2.75 : 0.177 : 0.246$	TSSG	R32	$Pr_{0.94}Y_{0.09}Sc_{2.97}(BO_3)_4$ (SEM/EDX) (periphery part) / $Pr_{0.93}Y_{0.10}Sc_{2.96}(BO_3)_4$ (SEM/EDX) (central part)	9.8256(5) 7.9038(4) 9.8179(6) 7.9029(1)	[113]
$NdSc_3(BO_3)_4$–$\alpha GdSc_3(BO_3)_4$ System					
$Nd_{1.125}Gd_{0.125}Sc_{2.75}(BO_3)_4$	Czochralski	P321	$(Nd_{0.8}Gd_{0.2(1)})Sc_3(BO_3)_4$ (XRD: P321 [3])		[35]
$NdSc_3(BO_3)_4$–$\alpha YSc_3(BO_3)_4$ System					
$Pr_2O_3 : Y_2O_3 : Sc_2O_3 : HBO_2 : Li_2(CO_3) : LiF = 0.125 : 0.125 : 0.5 : 1.94 : 0.221 : 0.205$	TSSG	R32	$Nd_{0.86}Y_{0.21}Sc_{2.93}(BO_3)_4$ (SEM/EDX) (periphery part) / $Nd_{0.87}Y_{0.18}Sc_{2.95}(BO_3)_4$ (SEM/EDX) (part under the seed)	9.7588(5) 7.9186(9) 9.7631(3)7.9210(7)	[113]
$CeSc_3(BO_3)_4$–$NdSc_3(BO_3)_4$–$GdSc_3(BO_3)_4$ System					
$(Ce_{0.65}Nd_{0.25}Gd_{0.10})Sc_3(BO_3)_4$	Czochralski	P321	$(Ce_{0.14(4)}Nd_{0.46}Gd_{0.13})Sc_3(BO_3)_4$ (XRD; R32 [3])		[35]
$Ce_{0.76}Nd_{0.30}Gd_{0.14}Sc_{2.8}(BO_3)_4$	Czochralski	P321	$(Ce_{0.57}Nd_{0.25}Gd_{0.18})Sc_3(BO_3)_4$ (XRD; P321 [3])		[35]

Table 2. *Cont.*

Initial Composition	Synthesis Method[1]	Space Group	Refined Composition (Method)[2]	Unit Cell Parameters, a, c, Å	Reference
		$CeSc_3(BO_3)_4$–$NdSc_3(BO_3)_4$–«$LuSc_3(BO_3)_4$»System			
$Ce_{1.2}Nd_{0.05}Lu_{0.3}Sc_{2.65}(BO_3)_4$	Czochralski	P321		21 reflections (space group R32): $a' = 7.895$, 9.776(6), 7.937(5); 18 reflections (space group A2): $a' = 7.895$, $b' = 9.749$, $c' = 16.817$, $\alpha = 90.23°$, $\beta = 89.95°$, $\gamma = 90.18°$	[35]
		$LaSc_3(BO_3)_4$–«$ErSc_3(BO_3)_4$»–«$YbSc_3(BO_3)_4$»System			
$LaYb_{0.15}Er_{0.015}Sc_{2.835}(BO_3)_{3.8}$, $LaYb_{0.3}Er_{0.015}Sc_{2.685}(BO_3)_{3.8}$, $LaYb_{0.3}Er_{0.03}Sc_{2.67}(BO_3)_{3.8}$, $LaYb_{0.36}Er_{0.015}Sc_{2.625}(BO_3)_{3.8}$	Czochralski	C2/c			[35]

[1] SSR—solid-state reaction; TSSG—top-seeded solution growth.[2] ICP—inductively coupled plasma elemental analysis; XRPD—X-ray powder diffraction; EDS—energy dispersive spectra standardless analysis; XRD—X-ray (single-crystal) diffraction; SEM/EDX—scanning electron microscopy with energy dispersive X-ray spectroscopy. [3] A structure and composition were refined in the space group specified.

3.2.1. LaSc$_3$(BO$_3$)$_4$-NdSc$_3$(BO$_3$)$_4$ System

The XRD study of the (La$_{1-x}$Nd$_x$)Sc$_3$(BO$_3$)$_4$ (LNSB) solid solutions (x = 0.0–0.3) grown by the Czochralski method showed their crystallization in the space group $C2/c$ [35,87]. The non-centrosymmetric trigonal phase $R32$ was discovered only in some crystals with high Nd^{3+} concentration ($x \geq 0.5$) [100]. A trigonal symmetry was found only for NSB crystal (Ir seed). For LSB (Ir seed) as well as (Nd$_{0.5}$La$_{0.5}$)Sc$_3$(BO$_3$)$_4$ and (Nd$_{0.1}$La$_{0.9}$)Sc$_3$(BO$_3$)$_4$ with the LSB seed, a monoclinic symmetry was established. It confirms a role of a seed in symmetry of grown crystals.

Symmetry of LNSB solid solutions (x = 0 to 1) obtained by the solid-state reaction and the flux method, was demonstrated by DTA, X-ray diffraction, second-harmonic generation, and fluorescence lifetime measurements [25]: for x = 1.0, i.e., for NdSc$_3$(BO$_3$)$_4$, only trigonal phase existed, and for $x \leq 0.5$, in the LNSB sample, two phases, with the space group $R32$ or $C2/c$, were obtained depended on the crystallized temperature (according to powder XRD, the phase with the space group $R32$ could be formed below 1100 °C). For the LSB (x = 0), the pure trigonal phase was difficult to obtain from the solid-state reaction. For example, in the LSB sample sintered at 1000 °C for 2 h, the huntite-family monoclinic phase and LaBO$_3$ were found.

3.2.2. LaSc$_3$(BO$_3$)$_4$ -«GdSc$_3$(BO$_3$)$_4$» System

Polycrystalline La$_x$Gd$_{1-x}$Sc$_3$(BO$_3$)$_4$ (LGSB) solid solutions were obtained by the conventional solid-state reaction [101–104], and single-crystal ones were grown by the high-temperature TSSG method using the Li$_6$B$_4$O$_9$–LiF as a flux [101,102] and Czochralski method [103,104] (Table 2). A crystallization of polycrystalline samples in the space group $R32$ for $0.20 \leq x \leq 0.80$ was established [101,102]. According to the XRD analysis, a single crystal grown by the high-temperature TSSG method crystallizes in the space group $R32$ and have the composition La$_{0.78}$Gd$_{0.22}$Sc$_3$(BO$_3$)$_4$, refined using a full-matrix least-squares refinement on F^2 (SHXLXL-97) using the occupancy factor for the La site fixed to a value that was determined from the ICP elemental analysis [101]. The composition of the single crystal grown by the Czochralski method from the charge La$_{0.678}$Gd$_{0.572}$Sc$_{2.75}$(BO$_3$)$_4$ is determined by the inductively coupled plasma atomic emission spectrophotometry (ICP-AES) method to be La$_{0.64}$Gd$_{0.41}$Sc$_{2.95}$(BO$_3$)$_4$ (space group $R32$) [103]. In addition, the LGSB single crystals were obtained by the Czochralski method from the charge with initial composition La$_{0.75}$Gd$_{0.5}$Sc$_{2.75}$(BO$_3$)$_4$ without refinement of crystal real composition [104].

According to [101], as a result of investigation of polycrystalline LGSB samples, it was revealed that the stoichiometry of the single trigonal phase followed the relationship of Sc/(Gd + La) = 3 and the deviations from these values afford additional peaks of impurities in the X-ray powder diffraction patterns. Therefore, Gd occupies the La site rather than Sc site [100]. It is confirmed by the Gheorghe et al. [103]: according to the ICP-AES method, the stoichiometry of Sc is close to 3. According to [103,104], for polycrystalline La$_{1-x}$Gd$_x$Sc$_3$(BO$_3$)$_4$ solid solutions, it was established that the structural changes from monoclinic (space group $C2/c$) to trigonal (space group $R32$) is complete for a Gd content larger than 0.3. The XRD investigation of a series of single crystal LGSB solid solutions, grown in 3:1 wt/wt LiBO$_2$/La$_{1-x}$Gd$_x$Sc$_3$(BO$_3$)$_4$ solvents or a 20 wt% CaF$_2$ flux, showed that the $R32$ structural integrity is maintained for $0.3 \leq x \leq 0.7$ [88].

3.2.3. LaSc$_3$(BO$_3$)$_4$ -«YSc$_3$(BO$_3$)$_4$» System

Polycrystalline and single-crystal La$_x$Y$_{1-x}$Sc$_3$(BO$_3$)$_4$ (LYSB) solid solutions were obtained by the solid-state reaction [31,102,105] and the high-temperature TSSG method using a lithium-borate flux [31,102,105–108], respectively (Table 2). An investigation of phase equilibria in the LaBO$_3$-ScBO$_3$-YBO$_3$ system [31,102,105] allowed to establish the limits of existence of huntite-type trigonal (space group $R32$) Y$_x$La$_y$Sc$_z$(BO$_3$)$_4$ solid solutions: $0.29 < x < 0.67$, $0.67 < y < 0.82$, and $2.64 < z < 3.00$. Selected initial compositions of samples within the trigonal-huntite region of the phase diagram are given in [31,102,105] (Table 2). The space group $R32$ has been determined for these samples by the powder XRD, and then was proven on the basis of the X-ray single-crystal refinement of structure of $La_{0.72}Y_{0.57}Sc_{2.71}(BO_3)_4$. Occupancy factors for the La and Sc sites were fixed to the values that were determined by ICP elemental analysis, as deviations from these occupancies afforded increased residuals in the least-squares refinements. The solid solution with the trigonal symmetry, grown by the TSSG method, had an average composition $La_{0.826}Y_{0.334}Sc_{2.84}(BO_3)_4$, determined by the ICP-AES [109].

According to the refinement of atomic coordinates and displacement parameters of LYSB structure of single crystal using a full-matrix least squares refinement on F^2 (SHELXL-97) [31], its composition can be written as $La_{0.72}Y_{0.57}Sc_{2.71}(BO_3)_4$ (space group $R32$), occupancy factors for the La, Y and Sc sites being fixed to the values determined by ICP-AES elemental analysis. Based on the comparison of cell volumes for the compositions $La_{0.77}Y_{0.28}Sc_{2.95}(BO_3)_4$ and $Y_{0.47}La_{0.75}Sc_{2.78}(BO_3)_4$ (La, Y, and Sc occupancies were determined by ICP-AES elemental analysis), in which the La concentrations are similar, the higher Y concentration of $La_{0.75}Y_{0.47}Sc_{2.78}(BO_3)_4$ ($V = 670.5(1)$ $Å^3$) relative to that of $La_{0.77}Y_{0.28}Sc_{2.95}(BO_3)_4$ ($V = 666.35(9)$ $Å^3$) leads to a 0.6% increase in cell volume, which can only occur if the Y atom substitutes the small Sc site [31]. The structure of $La_{0.75}Y_{0.25}Sc_3(BO_3)_4$ single crystal (space group $R32$) grown by the flux method was determined [108]. After full-matrix refinement with isotropic displacement coefficients on each atom, the occupancies of the La and Sc sites were refined. No significant change in the Sc occupancy factor was observed, so it was subsequently fixed to unity. Occupancy of the La site was significantly reduced, indicating occupation of Y on the La site. On the basis of the systematic condition $-h + k + l = 3n$ for the *hkil*, the statistical analysis of the intensity distribution, packing considerations, and the successful solution and refinement of the structure, the crystal was found to form in the non-centrosymmetric space group $R32$.

3.2.4. $LaSc_3(BO_3)_4$ -«$ErSc_3(BO_3)_4$» System

Durmanov et al. [35] selected the optimal compositions of charge with a B_2O_3 deficiency, which prevents its deposition on the surface of the growing crystal, since liquid B_2O_3 drains into the high temperature region and dissolves the growing crystal. As a result, optically qualitative activated $LaEr_{0.006}Sc_{2.994}(BO_3)_{3.8}$ and $LaEr_{0.015}Sc_{2.985}(BO_3)_{3.8}$ crystals with the space group $C2/c$ were obtained by the Czochralski method. Based on the crystallochemical similarity of Er with the Lu and Tb atoms, which, according to the XRD investigations, enter the Sc site [87], Durmanov et al. [35] suggested that the Er^{3+} ions also replace the Sc^{3+} ones. An increase in the concentration of Er^{3+} ions in the initial melt above 12 at % can lead to cracking of crystals and a change in their symmetry.

3.2.5. $LaSc_3(BO_3)_4$ -«$YbSc_3(BO_3)_4$» System

The optimal compositions of the charge with the lack of B_2O_3 to grow optically-qualitative crystals with the space group $C2/c$ by the Czochralski method are $LaYb_{0.15}Sc_{2.85}(BO_3)_{3.8}$, $LaYb_{0.3}Sc_{2.7}(BO_3)_{3.8}$, $LaYb_{0.36}Sc_{2.64}(BO_3)_{3.8}$ [35]. Durmanov et al. [35] suggested that the Yb^{3+} ions enter the Sc site based on the crystallochemical similarity of Yb with the Lu and Tb, which, according to the XRD investigations, occupy the Sc sites [87]. The maximum allowable concentration of Yb^{3+} ions to grow laser crystals is 10 at % (1.3×10^{21} cm^{-3}); higher concentration cannot be achieved using the growth technology described in [35].

3.2.6. $LaSc_3(BO_3)_4$ -«$LuSc_3(BO_3)_4$» System

Polycrystalline and single-crystal solid solutions with the initial composition $La_xLu_ySc_z(BO_3)_4$ ($x + y + z = 4$) were obtained by the solid-state reaction and the high-temperature TSSG method with a $Li_6B_4O_9$ flux [110], respectively (Table 2). According to single-crystal XRD measurements, the solid solution having the composition $La_{0.95}Lu_{0.66}Sc_{2.39}(BO_3)_4$, determined from the ICP elemental analysis, has been found to crystallize in the space group $R32$. Li et al. [110] suggested that the Lu atoms replace not only the La atoms in the trigonal prisms but also the Sc atoms in the octahedra; the formula for demonstrating Lu-atom substitution can be written as $(Lu_{0.05}La_{0.95})(Lu_{0.61}Sc_{2.39})(BO_3)_4$.

3.2.7. LaSc$_3$(BO$_3$)$_4$ -«BiSc$_3$(BO$_3$)$_4$» System

Solid solutions with the general composition La$_x$Bi$_y$Sc$_z$(BO$_3$)$_4$ were obtained in the form of polycrystals and single crystals using the solid-state reaction [111] and the spontaneous crystallization method using the Bi$_2$O$_3$–B$_2$O$_3$ as a flux [111,112], respectively (Table 2). Xu et al. [111] established that the single trigonal phase covers the composition range of $0.27 < x < 0.52$, $0.67 < y < 0.82$, and $2.74 < z < 2.95$. According to Xu et. al. [111], the Bi atoms replace not only the La atoms in the trigonal prisms but also the Sc atoms in the octahedra, and, hence, the general composition can be written as (La$_{1-x}$Bi$_x$)(Bi$_y$Sc$_{3-y}$)(BO$_3$)$_4$. The unit-cell volume (*V*) varies with the sizes of both the trigonal prism and the octahedron, and it can be expressed in the equation by using atomic radii (*R*) and appropriately weighted occupancies for the Bi, La, and Sc atoms. Based on the equation given in [111], an approximately linear relationship between the effective concentration (*Ceff*) and *V* was obtained, which indicates that the Bi atoms do indeed appear to be distributed across both sites. For the single crystal with the composition La$_{0.82}$Bi$_{0.27}$Sc$_{2.91}$(BO$_3$)$_4$, determined from the ICP-AES elemental analysis, the space group *R*32 was found using the XRD analysis [111] (Table 2).

Single-crystal La$_{0.91}$Bi$_{0.21}$Sc$_{2.88}$(BO$_3$)$_4$ solid solution, obtained by the spontaneous crystallization, the composition of which was checked with the energy dispersive spectra standardless analysis, has the trigonal structure with the space group R32, atomic sites and isotropic displacement factors being refined without any refinement of site occupancies [112] (Table 2).

3.2.8. CeSc$_3$(BO$_3$)$_4$-NdSc$_3$(BO$_3$)$_4$ System

Single-crystal solid solution with the composition of the initial charge (Ce$_{0.53}$Nd$_{0.64}$Sc$_{2.83}$)(BO$_3$)$_4$, grown by the Czochralski method, crystallizes in the space group *P*321 [35]. A comparison of the unit cell parameters, atomic coordinates, and interatomic distances of this phase with those for the NdSc$_3$(BO$_3$)$_4$ allowed to receive the composition (Ce$_{0.5}$Nd$_{0.5(1)}$)Sc$_3$(BO$_3$)$_4$. The structure disordering of (Ce$_{0.5}$Nd$_{0.5(1)}$)Sc$_3$(BO$_3$)$_4$ is more pronounced compared with the NdSc$_3$(BO$_3$)$_4$ [35].

3.2.9. CeSc$_3$(BO$_3$)$_4$-«GdSc$_3$(BO$_3$)$_4$» System

Single-crystal solid solutions obtained by the Czochralski method from the charges having compositions (Ce$_{0.8}$Gd$_{0.2}$)Sc$_3$(BO$_3$)$_4$, (Ce$_{0.7}$Gd$_{0.3}$)Sc$_3$(BO$_3$)$_4$, and Ce$_{0.9}$Gd$_{0.35}$Sc$_{2.75}$(BO$_3$)$_4$ crystallize in the space group *P*321 (Table 2). According to the XRD analysis of microcrystals, 67% of diffraction reflections do not obey the extinction laws of the space group *R*32, but they are indexed in the space group *P*321, i.e., in the superstructure relative to the huntite structure. A refinement of the crystal structure of solid solutions with the nominal compositions (Ce$_{0.80}$Gd$_{0.20}$)Sc$_3$(BO$_3$)$_4$ and Ce$_{0.9}$Gd$_{0.35}$Sc$_{2.75}$(BO$_3$)$_4$ based on the strongest reflections (33%) in the subcell with the space group *R*32, i.e., in the huntite structure (atomic coordinates, atomic displacement parameters, occupancies of all crystallographic sites except for B and O ones), showed that their compositions can be described as (Ce$_{0.803(2)}$Gd$_{0.091(4)}$Sc$_{0.106(4)}$)Sc$_3$(BO$_3$)$_4$ (*R* = 3.04 %) and (Ce$_{0.949(1)}$Gd$_{0.024(1)}$Sc$_{0.027(1)}$)Sc$_3$(BO$_3$)$_4$ (*R* = 1.93 %), respectively. The refined crystal compositions correlate with the unit cell parameters (r$_{Ce}$>r$_{Gd}$>r$_{Sc}$), which are indicators of a composition. A refinement of the structure of the microcrystal with the nominal composition (Ce$_{0.80}$Gd$_{0.20}$)Sc$_3$(BO$_3$)$_4$ in the space group *P*321 (i.e., taking into account all diffraction reflections, both strong and weak, but with the $I \geq 3\sigma(I)$; *R* = 3.13%) allowed writing the composition as (Ce$_{0.485(3)}$Gd$_{0.009}$Sc$_{0.006}$)(Ce$_{0.465(6)}$Gd$_{0.017}$Sc$_{0.018}$)Sc$_3$(BO$_3$)$_4$ or (Ce$_{0.950(3)}$Gd$_{0.026}$Sc$_{0.024}$)Sc$_3$(BO$_3$)$_4$ (Tables 3 and 4).

Table 3. Coordinates of atoms, atom displacement parameters ($B_{eq} \times 10^2$, Å^2), and site occupancies (p) in the structure of $(Ce_{0.80}Gd_{0.20})Sc_3(BO_3)_4$ solid solution in the space group $R32$ and $P321$ according to the XRD data (AgKα).

Parameter	Space Group $R32$	Space Group $P321$	Parameter	Space Group $R32$	Space Group $P321$	Parameter	Space Group $R32$	Space Group $P321$
Z	3		μ, mm^{-1}	3.90	3.87	θ_{max}, deg	25.865	
a, Å	9.781(1)		Refl. read/unique ($I > 2\sigma(I)$)	1765/591	1765/1765	wR_2	0.0500	0.0753
c, Å	7.948(1)		No. of parameters	37	90	R_1 ($I > 2\sigma(I)$)	0.0304	0.0313
V, Å^3	658.53					S	0.857	0.978
Site	Ce1/Gd1/Sc	Ce1/Gd1/Sc	Site	Sc1	Sc1	Site	O1	O1
x	0	0	x	0.2141(1)	0.50053(9)	x	0.0193(4)	0.4560(3)
y	0	0	y	1/3	0	y	0.2116(4)	0.1408(3)
z	0	0	z	1/3	0	z	0.1813(4)	0.5228(3)
B_{eq}	1.06(1)	1.11(1)	B_{eq}	0.92(3)	0.92(2)	B_{eq}	1.58(6)	1.44(5)
p(Ce1/Gd1/Sc1)	0.1338(3)/0.0152(7)/0.0177(7)	0.1617(9)/0.0030(9)/0.0020(9)	p	0.5	0.5	p	1.0	1.0
Site	Ce2/Gd2/Sc	Ce2/Gd2/Sc	Site		Sc2	Site	O2	O2
x		1/3	x		0.21195(9)	x	0.58777	0.1930(4)
y		2/3	y		0.33235(6)	y	0.5	−0.0209(3)
z		0.66638(4)	z		0.32954(6)	z	0.5	0.1769(3)
B_{eq}		1.04(1)	B_{eq}		0.86(2)	B_{eq}	2.4(1)	1.51(5)
p(Ce2/Gd2/Sc2)		0.155(2)/0.006(2)/0.006(2)	p		1.0	p	0.5	1.0
			Site	B1	B1	Site	O3	O3
			x	0.4516(6)	0.4536(5)	x	−0.1399(5)	0.3114(3)
			y	0	0	y	0	0.2479(4)
			z	0.5	0.5	z	0.5	0.1754(3)
			B_{eq}	1.30(8)	1.05(8)	B_{eq}	1.54(8)	1.96(8)
			p	0.16667	0.5	p	0.5	1.0
			Site	B2	B2	Site		O4
			x	0	0.3282(4)	x		0.4755(4)
			y	0	0.1146(5)	y		0.1243(3)
			z	0.5	0.1705(5)	z		0.1562(3)
			B_{eq}	1.3(1)	1.28(6)	B_{eq}		1.56(5)
			p	0.16667	1.0	p		1.0
			Site		B3	Site		O5
			x		0	x		0.5989(5)
			y		0	y		0
			z		0.5	z		0.5
			B_{eq}		1.4(2)	B_{eq}		2.02(8)
			p		0.16667	p		0.5
			Site		B4	Site		O6
			x		2/3	x		0.1405(4)
			y		1/3	y		0.1405(4)
			z		0.84736(6)	z		0.5
			B_{eq}		0.82(9)	B_{eq}		1.22(4)
			p		0.3333	p		0.5
						Site		O7
						x		0.6745(3)
						y		0.1983(3)
						z		−0.1536(2)
						B_{eq}		1.22(4)
						p		1.0

Table 4. Main interatomic distances in the structure of of $(Ce_{0.80}Gd_{0.20})Sc_3(BO_3)_4$ solid solution in the space group $R32$ and $P321$ according to the XRD data $(AgK\alpha)$.

Parameter	Space Group $R32$	Parameter	Space Group $P321$
Ce1/Gd1/Sc		Ce1/Gd1/Sc	
$-6 \times$ O1	2.450(3)	$-6 \times$ O2	2.443(3)
		Ce2/Gd2/Sc	
		$-3 \times$ O4	2.417(3)
		$-3 \times$ O1	2.484(3)
		[Ce2/Gd2/Sc-O]$_{av.}$	2.4505
Sc1		Sc1	
$-2 \times$ O1	2.059(3)	$-2 \times$ O7	2.091(3)
$-2 \times$ O3	2.118(3)	$-2 \times$ O4	2.110(3)
$-2 \times$ O2	2.148(4)	$-2 \times$ O3	2.211(3)
[Sc1-O]$_{av.}$	2.129	[Sc1-O]$_{av.}$	2.134
		Sc2	
		$-1 \times$ O2	2.035(3)
		$-1 \times$ O1	2.041(3)
		$-1 \times$ O5	2.113(3)
		$-1 \times$ O3	2.124(3)
		$-1 \times$ O6	2.129(2)
		$-1 \times$ O7	2.149(2)
		[Sc2-O]$_{av.}$	2.0985
B1		B1	
$-1 \times$ O2	1.33(1)	$-2 \times$ O1	1.378(4)
$-2 \times$ O1	1.368(5)	$-1 \times$ O5	1.412(8)
	1.355	[B1-O]$_{av.}$	1.389
		B2	
		$-1 \times$ O3	1.289(6)
		$-1 \times$ O2	1.325(5)
		$-1 \times$ O4	1.401(5)
		[B2-O]$_{av.}$	1.338
B2		B3	
$-3 \times$ O3	1.369(5)	$-3 \times$ O6	1.375(4)
		B4	
		$-3 \times$ O7	1.360(3)

3.2.10. $CeSc_3(BO_3)_4$-«$YSc_3(BO_3)_4$» System

A refinement of the crystal structure of the Czochralski-grown powdered crystal with the initial composition $Ce_{1.25}Y_{0.3}Sc_{2.45}(BO_3)_4$ by the Rietveld method in the huntite subcell with the space group $R32$ (the proper space group is $P321$: 67% of diffraction reflections do not obey the extinction laws of the space group $R32$) showed its real composition as $(Ce_{0.78}Y_{0.22(2)})Sc_3(BO_3)_4$ [35]. A decrease in the Ce site occupancy was observed (the form factor or atomic factor is proportional to the atomic number), which indicates a partial replacement of the Ce atoms by the Y ones, and the presence of Sc atoms in this site has not been considered. The XRD analysis of the cation sites, except for the B site, in the structure of micropart of the same crystal (space group $R32$) resulted in the composition $(Ce_{0.995(1)}Y_{0.002(1)}Sc_{0.003(1)})Sc_3(BO_3)_4$ ($R = 7.73$ %), i.e., with the absence of Y atoms in the Sc site. Different compositions refined for powder and single-crystal samples can be explained by the heterogeneity of bulk crystal composition.

3.2.11. $CeSc_3(BO_3)_4$-«$LuSc_3(BO_3)_4$» System

A refinement of structure of the powdered sample in the space group $R32$ (the proper space group is $P321$) using the full-profile method showed that the composition of the Czochralski-grown crystal can be written as $Ce(Lu_{0.17(1)}Sc_{2.83})(BO_3)_4$ with a presence of the Lu ions in the octahedral sites of the structure together with the Sc ones [35]. The XRD analysis of the structure of micropart of the same crystal confirmed the presence of Lu atoms in the octahedral site, but with a lower content: $Ce(Sc_{2.910(30)}Lu_{0.090})(BO_3)_4$ ($R = 5.95\%$).

3.2.12. $PrSc_3(BO_3)_4$ -$NdSc_3(BO_3)_4$ System

A solid solution with the charge composition $(Pr_{0.99}Nd_{0.11}Sc_{2.9})(BO_3)_4$ grown by the Czochralski method crystallizes in the space group $P321$ [83].

3.2.13. $PrSc_3(BO_3)_4$-«$YSc_3(BO_3)_4$» System

A solid solution with the general composition $Pr_xY_ySc_z(BO_3)_4$ grown by the TSSG method using eutectic $LiBO_2$–LiF as a flux crystallizes in the space group $R32$, as determined by the XRD on the powder sample [113] (Table 2). According to scanning electron microscopy with energy dispersive X-ray spectroscopy (SEM/EDX), a peripheral part of the crystal has the composition $Pr_{0.94}Y_{0.09}Sc_{2.97}(BO_3)_4$, while the composition of the central part of the crystal is $Pr_{0.93}Y_{0.10}Sc_{2.96}(BO_3)_4$ [113]. It should be noted that the unit cell parameters of samples from two parts of the crystal with almost identical compositions are significantly different, although there is a tendency for the parameters to increase with increasing Pr content ($r_{Pr} > r_Y$) (Table 2).

3.2.14. $NdSc_3(BO_3)_4$-«$GdSc_3(BO_3)_4$» System

The XRD analysis of the Czochralski-grown crystals with the initial compositions $Nd_{1.125}Gd_{0.125}Sc_{2.75}(BO_3)_4$ (the refined composition $(Nd_{0.8}Gd_{0.2(1)})Sc_3(BO_3)_4$ was evaluated by comparing the structural parameters of this phase with those for the $NdSc_3(BO_3)_4$), $Nd_{1.04}Gd_{0.26}Sc_{2.7}(BO_3)_4$, and $Nd_{0.91}Gd_{0.39}Sc_{2.7}(BO_3)_4$ showed their crystallization in the space group $P321$ [35].

3.2.15. $NdSc_3(BO_3)_4$-«$YSc_3(BO_3)_4$» System

Single-crystal solid solution with the charge composition $Nd_xY_ySc_z(BO_3)_4$ was grown by the TSSG method [113]. The space group $R32$ was established by the powder XRD method. According to the SEM/EDX, a periphery part of the crystal has the composition $Nd_{0.86}Y_{0.21}Sc_{2.93}(BO_3)_4$, while the region under the seed is $Nd_{0.87}Y_{0.18}Sc_{2.95}(BO_3)_4$. A situation is similar to that found for solid solutions in the $PrSc_3(BO_3)_4$-«$YSc_3(BO_3)_4$» system: significant differences in the unit cell parameters of samples taken from different parts of the crystal with almost identical compositions (taking into account a measurement error) are found, a noticeable correlation between the unit cell parameters and the Y content ($r_{Nd} > r_Y$) being observed.

3.2.16. $CeSc_3(BO_3)_4$-$NdSc_3(BO_3)_4$-«$GdSc_3(BO_3)_4$» System

The XRD study of a solid solution grown by the Czochralski method showed that its structure has the huntite subcell with the space group $R32$ (Table 2). In this case, a limited number of weak diffraction reflections go into a superstructural trigonal cell with parameters doubled with respect to the huntite ones: $A = 2a_{R32}$, $C = 2c_{R32}$ [35]. The single-crystal XRD analysis of a solid solution with the charge composition $Ce_{0.76}Nd_{0.30}Gd_{0.14}Sc_{2.8}(BO_3)_4$ and estimated composition $(Ce_{0.57}Nd_{0.25}Gd_{0.18})Sc_3(BO_3)_4$ revealed a significant number of superstructure reflections with the space group $P321$ [35]. Durmanov et al. [35] believe that the choice of the initial melt composition and growth conditions may rule out a formation of superstructure in the $(Ce,Nd,Gd)Sc_3(BO_3)_4$ solid solution; among these compositions may be found a congruent melting one.

3.2.17. $CeSc_3(BO_3)_4$-$NdSc_3(BO_3)_4$-«$LuSc_3(BO_3)_4$» System

A large number of maximally-split reflections was revealed as a result of XRD investigation of a small chip of the Czochralski-grown crystal with the initial composition $Ce_{1.2}Nd_{0.05}Lu_{0.3}Sc_{2.45}(BO_3)_4$.

The unit cell parameters determined by an auto-indexing the most intense 21 reflections correspond to the huntite cell with the parameters $a_{R32} = 9.776(6)$, $c_{R32} = 7.937(5)$ Å [35]. In the interval of interplanar distances $d = 2.32$–3.14 Å, weak and diffuse reflections, along with the strong ones, were found. The unit cell parameters determined from 18 reflections in the same interval of interplanar distances were found to be $a' = 7.895$, $b' = 9.749$, $c' = 16.817$Å, $\alpha = 90.23°$, $\beta = 89.95°$, $\gamma = 90.18°$. The obtained triclinic unit cell is a pseudo-monoclinic one (space group $A2$) with the parameters correlated with those of the huntite cell: $a' = c_{R32}$, $b' = b_{R32}$, $c' = 2a_{R32}\cos 30°$, $\alpha \sim 90°$, $\beta \sim 90°$, $\gamma \sim 90°$. However, two weak (the intensities are 44 and 65 times less than the intensity of the strongest reflection in the above-mentioned interval) and diffuse (peak width are $1.56°$ and $2.73°$) reflections were not indexed with these parameters. Taking into account one weak (44 times weaker than the strongest reflection) and diffuse (width is $1.56°$) reflection, the a' unit cell parameter doubles: $a'' = 2a'$, $b'' = b'$, $c'' = c'$. However, the second reflection remained non-indexed with these parameters. A superstructure with the unit cell parameters a', b', c' was also found for the solid solution with the composition $(Ce_{0.80}Gd_{0.20})Sc_3(BO_3)_4$ [84].

3.2.18. LaSc$_3$(BO$_3$)$_4$-«ErSc$_3$(BO$_3$)$_4$»-«YbSc$_3$(BO$_3$)$_4$» System

Durmanov et al. [35] selected the optimal compositions of the melt to obtain crystals by the Czochralski method (Table 2). According to the XRD analysis, all grown crystals have a monoclinic symmetry, probably the space group $C2/c$. Based on the previous studies [76], Durmanov et al. [35] suggested that the Er^{3+} and Yb^{3+} ions replace the Sc^{3+} ones, and their maximum concentration in the initial melt should not exceed 12 at %, since a higher content may affect the cracking of crystals and change the symmetry (similar to solid solutions in the LaSc$_3$(BO$_3$)$_4$ – «ErSc$_3$(BO$_3$)$_4$» system).

Thus, huntite-family solid solutions of rare-earth scandium borates crystallize either in a monoclinic symmetry with the centrosymmetric space group $C2/c$ (in many works, the centrosymmetry is not confirmed) or in a trigonal one with the non-centrosymmetric space groups $R32$ (requires confirmation) or $P321$. In most cases, due to methodological difficulties of the XRD experiment, the site occupancies were not refined for the structures under investigation, especially when several cations with close atomic scattering factors occupy one crystallographic site. However, an elemental analysis and a crystallochemical approach made it possible to suggest the most likely composition from different compositions of solid solutions.

4. Crystallochemical Features of the Huntite Family

The variety of compositions and different symmetries found for the huntite-family compounds and solid solutions allows to make their classification and systematization based on crystallochemical phenomena such as isomorphism (an ability of a system to form equivalents), morphotropy (a change in a crystal structure for a regular series of compounds having similar formula composition), polymorphism (an adaptability of a structure to external influences), and the adjacent polytypism (an ability of a substance to crystallize in various modifications with different packing of structural elements along one axis). Surely, there are no clear boundaries between them, but in the first approximation, this approach allows identifying general structural features and finding fundamental differences in the group of compounds or solid solutions.

In addition to the possibility of forming continuous or limited solid solutions (the unlimited (perfect) and limited (imperfect) isomorphism) and internal ones (preferably for rare-earth scandium borates), a morphotropy is characteristic of the huntite-family compounds. The main reason for the morphotropy, in this case, is the size factor, namely, a change in the ionic radius of the Ln cation (LnAl$_3$(BO$_3$)$_4$–LnGa$_3$(BO$_3$)$_4$–LnSc$_3$(BO$_3$)$_4$) (Figure 1) or M cation (LaAl$_3$(BO$_3$)$_4$–LaFe$_3$(BO$_3$)$_4$–LaCr$_3$(BO$_3$)$_4$–LaSc$_3$(BO$_3$)$_4$, TbAl$_3$(BO$_3$)$_4$–TbFe$_3$(BO$_3$)$_4$–TbCr$_3$(BO$_3$)$_4$–TbGa$_3$(BO$_3$)$_4$–TbSc$_3$(BO$_3$)$_4$, YbAl$_3$(BO$_3$)$_4$–YbFe$_3$(BO$_3$)$_4$–YbCr$_3$(BO$_3$)$_4$–YbGa$_3$(BO$_3$)$_4$) (Figure 11) [114]). *The morphotropic series* include nominally-pure and activated samples, for example, LaSc$_3$(BO$_3$)$_4$ (space group $C2/c$) – LaSc$_3$(BO$_3$)$_4$:Cr^{3+} (space group $P1$ or $P\bar{1}$) (an electronic structure factor) [76] and LaSc$_3$(BO$_3$)$_4$ (space group $C2/c$) – LaSc$_3$(BO$_3$)$_4$:Nd^{3+} (space group $C2$) (a size factor) [24].

The polymorphic transition from the space group $P3_121$ to the $R32$ one (from a low-symmetric to a high-symmetric modification) with increasing temperature is known for the $LnFe_3(BO_3)_4$ with the Ln = Eu–Er, Y [53,56–58,60,62–68] (Figures 1 and 13).

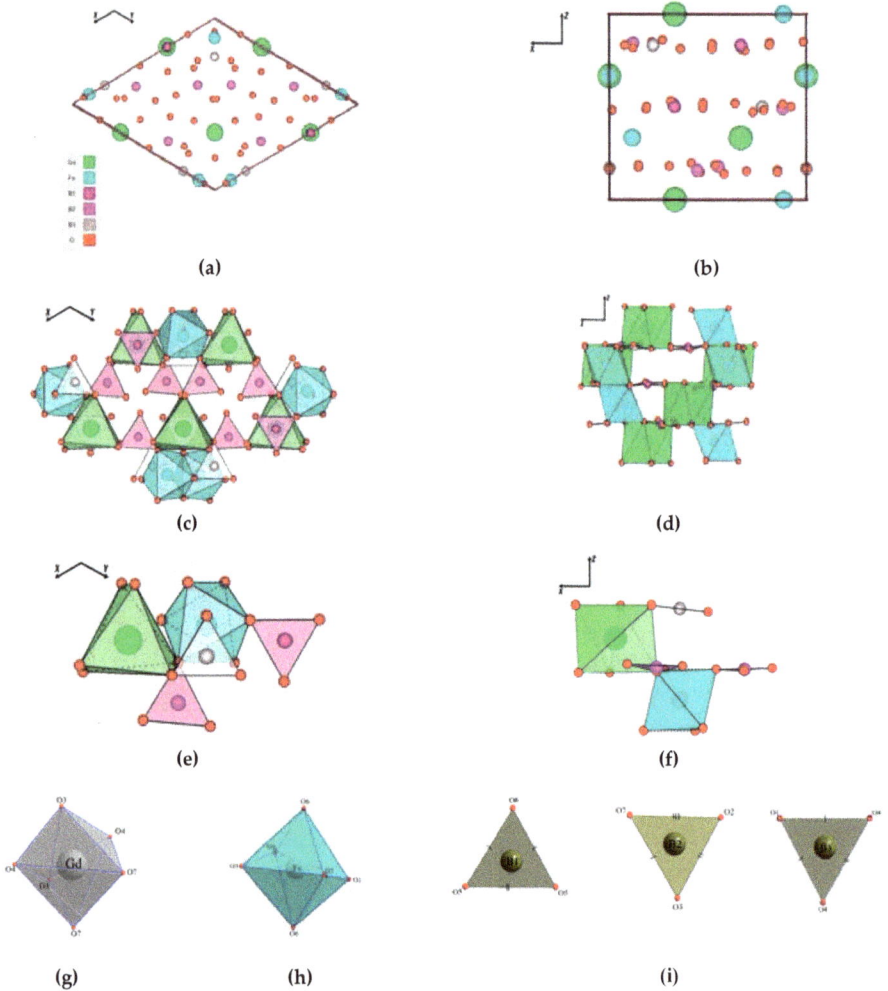

(a) (b) (c) (d) (e) (f) (g) (h) (i)

Figure 13. The unit cell of the $GdFe_3(BO_3)_4$ structure (space group $P3_121$) projected onto the (**a**) XY and (**b**) XZ planes; Combination of the coordination polyhedra projected onto the (**c**) XY and (**d**) XZ planes; Combination of selected coordination polyhedra projected onto the (**e**) XY and (**f**) XZ planes; Coordination polyhedra for the (**g**) Gd, (**h**) Fe, (**i**) B1–B3.

In the $LnFe_3(BO_3)_4$ structure with the low-temperature modification (space group $P3_121$) with the unit cell parameters (Figure 13a,b) similar to those for the huntite (space group $R32$) ($a = 9.5305$, $c = 7.5479$ Å for the $GdFe_3(BO_3)_4$), the polyhedra (Figure 13c–i) are the same as those in the structures of huntite-like compounds: a trigonal prism for the Ln (Figure 13g), an octahedron for the Fe (Figure 13h), and three triangles for the B atoms (in the huntite structure, there are two crystallochemically-different B atoms) (Figure 13i). A comparison of the XZ projections of the structures with the space groups $R32$ (Figure 3b) and $P3_121$ (Figure 13b) shows a similar alternation of atoms along the Z axis (Ln, Sc–B,

O–Ln, Sc–B, O) and polyhedra (Figure 3d,f and Figure 13d,f), but with corresponding gaps in the structure with the space group $P3_121$ (compare Figures 3c and 13c and Figures 3d and 13d) due to the absence of the second helical axis.

The space group $P3_121$ was derived for the $GdFe_3(BO_3)_4$ structure (90 K) from the systematic extinctions and was discriminated from other candidate space groups that comply with the same extinction conditions during the structure determination process; the polarity of the structure actually chosen (space group $P3_121$) was determined by Flack's x refinement [63]. Using circularly polarized X rays at the Dy L3 and Fe K absorption edges, Nakajima et al. [115] established that the single-crystal $DyFe_3(BO_3)_4$, which has the chiral helix structure of Dy and Fe ions on the screw axes, belongs to the left-handed space group $P3_221$, and this is in accord with the results on soft x-ray diffraction at Dy M5 absorption edge [116].

With an appropriate choice of the origin of the coordinates for the $LnFe_3(BO_3)_4$ unit cell (on the Gd atom, Figure 13b), the topological similarity of the structures with the space groups R32 and $P3_121$ is observed. At the same time, the high-temperature modification is more symmetrical than the low-temperature one, which is typical of polymorphism [117].

For phases with the compositions $(Ln,Sc)Sc_3(BO_3)_4$ and $LnSc_3(BO_3)_4$, a polymorphic "order–disorder" phase transition (if the space group R32 is proven for these compounds) from the space group P321 (a disordered structure with a partial or full ordering of the Ln^{3+} and Sc^{3+} ions over two trigonal-prismatic sites and that of the Sc^{3+} ions over two octahedral sites) to the space group R32 (an ordered structure with a statistical arrangement of the Ln^{3+} and/or Sc^{3+} ions in one trigonal-prismatic site and that of the Sc^{3+} ions in one octahedral site) can be assumed with increasing temperature.

On the other hand, it is possible that a kinetic 'order–disorder' phase transition (see, for example, [118,119]) occurs for rare-earth scandium orthoborates, i.e., a partially ordered phase (space group P321) is formed in the stability region of the disordered phase (space group R32) (a positional 'ordering–disordering') under an influence of kinetic (growth) factors. Growth dissymmetrization, as a rule, affects local parts of a crystal (hence, no more than 67 % of diffraction reflections, which did not correspond to the space group R32, were found), i.e., there is a different correlation between unit cells with different symmetries (a peculiar volume defect).

A necessary condition for this kind of ordering in the structure is, first of all, the presence of sites jointly occupied by several crystallochemically-different atoms and their concentration. For example, an increase in the content of Yb activator in the scheelite-family $(Na_{0.5}Gd_{0.5})MoO_4$:Yb crystals grown by the Czochralski method leads to an increase in the degree of structure deviation from the space group $I4_1/a$ and an increase in the orthorhombic distortion of the resulting superstructure [120]. For the $(Na_{0.5}Gd_{0.5})MoO_4$: 10% Yb crystals, ~50% of reflections, which are not inherent in the scheelite centrosymmetric space group $I4_1/a$, but typical for the non-centrosymmetric space group $P\bar{4}$, was revealed by the XRD experiment [120]. It should be noted that only ~ 4% of such reflections was found for the $(Na_{0.5}Gd_{0.5})WO_4$: 10% Tm crystals $(r_{Gd} > r_{Tm} > r_{Yb})$, and $(Na_{0.5}Bi_{0.5})MoO_4$ crystallizes in the space group $I\bar{4}$ [121]. An ordering depends on the growth method and synthesis conditions (for the Czochralski method: crystallization, cooling, and annealing rates; annealing and quenching temperatures; growth atmosphere, etc.). Indeed, an increase in growth rates from 4 to 6 mm/hour reduces a degree of structure ordering for the $(Na_{0.5}Gd_{0.5})WO_4$ crystal. The cooling rate of the crystals has a similar effect: a decrease in the cooling rate contributes to the formation of ordered non-centrosymmetric $(Na_{0.5}Gd_{0.5})WO_4$ and $(Na_{0.5}Gd_{0.5})WO_4$:Yb crystals (space group $I\bar{4}$) [122]) Hence, a growth dissymmetrization is most pronounced in conditions close to equilibrium.

According to [46,123], for the $LnAl_3(BO_3)_4$ compounds, a modification with the space group R32 is formed at low temperatures (~880–900 °C), phases with the centrosymmetric $C2/c$ symmetry crystallize in the higher temperature region, up to 1040–1050 °C, and a further temperature increase leads to a formation of the most disordered and metastable non-centrosymmetric modification with the C2 symmetry. It can be seen that structurally disordered modifications with low symmetry are formed at elevated temperatures, which contradicts the rules of *polymorphism*, but it is characteristic of polytypism [124]. Inclusions of one polytype in another were revealed for the $LnAl_3(BO_3)_4$ with the Ln = Nd and Gd by the IR spectroscopy using a factor group analysis for vibrations of the B–O bond [45]. Moreover, the structure of the monoclinic $SmAl_3(BO_3)_4$ crystal contains significant fragments of the trigonal polytype, and the structure of the trigonal $NdAl_3(BO_3)_4$ have a high content of domains of the monoclinic polytype. However, this conclusion was not confirmed by structural studies.

The unit cell parameters of the huntite structure with the space group $R32$ (the hexagonal cell: $a_{R32} = b_{R32} \neq c_{R32}$) and those of the huntite-family structures are related as (Figure 14):

- $a_{C2/c} = 0.666[c^2_{R32} + (a_{R32}\cos30°)^2]^{1/2}$, $b_{C2/c} = b_{R32}$, $c_{C2/c} = 0.666\,[(2c_{R32})^2 + (a_{R32}\cos30°)^2]^{1/2}$ (space group $C2/c$: $a = 7.7297(3)$, $b = 9.8556(3)$, $c = 12.0532(5)$ Å, $\beta = 105.405(3)°$) [82];
- $a_{C2/c} = 0.666[c^2_{R32} + (a_{R32}\cos30°)^2]^{1/2}$, $b_{C2/c} = b_{R32}$, $c_{C2/c} = [(2.5c_{R32})^2 + (a_{R32}\cos30°)^2]^{1/2}$ (space group $C2/c$: $a = 7.227(3)$, $b = 9.315(3)$, $c = 21.688(3)$ Å, $\beta = 95.90(2)°$) [97];
- $a_{c2} = 0.666[c^2_{R32} + (a_{R32}\cos30°)^2]^{1/2}$, $b_{c2} = b_{R32}$, $c_2 = 1.333[(1.25c_{R32})^2 + (a_{R32}\cos30°)^2]^{1/2}$ (space group $C2$: $a = 7.262(3)$, $b = 9.315(3)$, $c = 16.184(8)$ Å, $\beta = 90.37°$) [47].

It follows that two unit cell parameters are the same for the huntite-like structures, and the third *c* parameter is correlated with the other two.

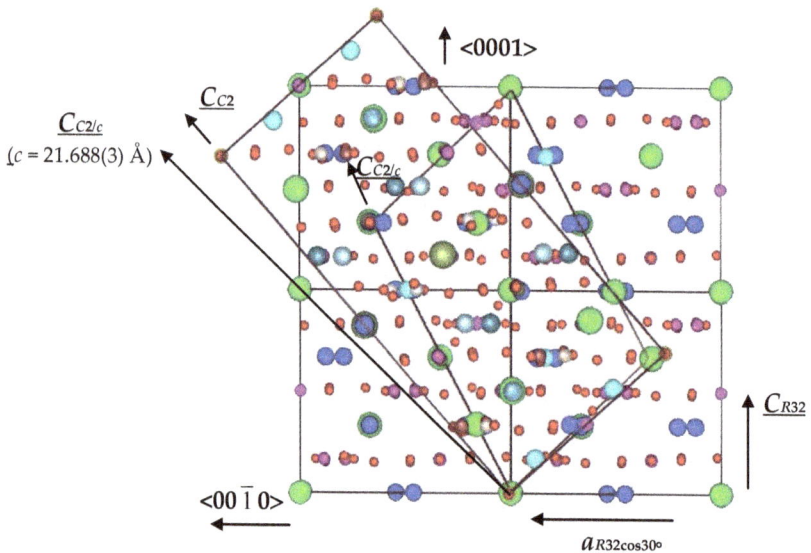

Figure 14. Genetic correlations between the unit cells for the huntite family compounds.

According to the results of the XRD analysis of the micropart of the crystal with the initial composition $Nd_{1.25}Sc_{2.75}(BO_3)_4$ (NSB–1.25), the unit cell parameters determined by auto-indexing of 21 reflections ($h0\bar{h}0$, $000l$) in the interval of interplanar distances $d = 2.01$–8.14 Å, correspond to the primitive trigonal cell with the c_{R32} parameter doubled with respect to the huntite one ($a = a_{R32} = 9.74$, $c = 2c_{R32} = 15.83$ Å) [89]. In the interval of interplanar distances $d = 3.96$–4.00 Å, several diffuse reflections with a width of 1.23–1.4° were found (the remaining reflections of approximately the same intensity had a width of 1.05°). Taking them into account (25 reflections), a primitive trigonal cell with the doubled a_{R32} and c_{R32} parameters ($A = 2a_{R32} = 19.526(3)$, $C = 2c_{R32} = 15.838(2)$ Å), as for the $(Ce_{0.41(4)}Nd_{0.46}Gd_{0.13})Sc_3(BO_3)_4$, was obtained. It should be noted that the refinement of the NSB–1.25 composition in the space group $R32$ by the Rietveld method allowed to find its composition as $NdSc_3(BO_3)_4$. An analysis of the atom displacement and positional parameters (first of all, for the B2 atoms) indicates a structure different from the huntite with the space group $R32$. A similar diffraction picture was observed for the solid solutions in $CeSc_3(BO_3)_4$ - $NdSc_3(BO_3)_4$ - «$GdSc_3(BO_3)_4$» and $CeSc_3(BO_3)_4$ - $NdSc_3(BO_3)_4$ - «$LuSc_3(BO_3)_4$» systems.

As a result of quenching of rare-earth aluminum orthoborates with the trigonal symmetry, in addition to the huntite-type modification (space group $R32$), a structural state with a full structural disorder in the alternation of layers along the c axis was revealed in the single-crystal diffraction patterns (diffuse bands on the reciprocal lattice along the c^* parameter) [46]. In the diffraction patterns of monoclinic phases (space groups $C2/c$ and $C2$), weak diffuse reflections, which should be absent due to extinction of space group reflections of the matrix structures, were also observed.

All the above-mentioned experimental facts, namely, a presence of diffuse areas along with the point ones in the diffraction patterns, a presence of strong diffraction reflections with a high symmetry, the similar a_{R32} and b_{R32} parameters and the c_{R32} parameter, which can be represented as a linear combination of vectors (for classical polytypes, the similar a and b parameters and the c one multiple to minimum) meet the general principles of polytypism and are optimally described in terms of OD (order–disorder) theory [125].

The main factors for the formation of polytypes for the $LnAl_3(BO_3)_4$ are a crystallization temperature (thermodynamic factor), a crystallization rate and a cooling rate in the flux method (kinetic factor), and a nature of the Ln cation and a r_{Ln}/r_{Al} ratio (crystallochemical factor) [27]. For the $LnAl_3(BO_3)_4$ with the Ln = Pr – Gd, two polytypic modifications with the space groups $R32$ and $C2/c$ occur: for the Ln = Pr and Nd, a monoclinic structure is more stable; for the Ln = Sm, Eu, and Gd, a trigonal modification is more stable, a monoclinic phase is formed in the flux at high temperatures and concentrations only. In the case of orthoborates with the Ln = Tb - Lu and Y, a modification with the huntite structure is only stable [123]. Beregi et al. [97] concluded that the starting crystallization temperature is a dominant factor in formation of the trigonal and monoclinic symmetry: for the phase with the nominal composition $Eu_{0.02}Tb_{0.12}Gd_{0.86}Al_3(BO_3)_4$, a higher (1080 °C) and lower (1060 °C) starting temperatures led to an appearance of modifications with the space groups $C2/c$ (a = 7.227(3), b = 9.315(3), c = 21.688(3) Å, β = 95.90(2)°) and $R32$ (a = 9.294(2), c = 7.251(2) Å), respectively.

Rare-earth scandium orthoborates, unlike other huntite-family rare-earth borates, exhibit a slightly different structural behavior. For example, for the $LaSc_3(BO_3)_4$, three modifications are known: a low-temperature (space group Cc, is a subgroup of the space group $C2/c$ and it is absent for other rare-earth orthoborates), a medium-temperature (space group $R32$; as was mentioned above, this group is denied by many researchers) and a high-temperature (space group $C2/c$) ones. The order of realization of the symmetry by the scandium borate crystals with increasing temperature is clearly different from that found for rare-earth aluminum borates: the most symmetrical structure crystallizes at high temperatures. It is quite possible that rare-earth scandium orthoborates are characterized primarily by polymorphs, although polytypes are also possible.

5. Conclusions

A critical analysis of the data on growth and structural diagnostics (composition, structure) of the huntite-family compounds and solid solutions indicates the most complete and consistent data obtained for rare-earth aluminum orthoborates. Based on the results of investigation of the $LnAl_3(BO_3)_4$ crystals with the Ln = Pr – Lu, Y obtained by the flux method (spontaneous crystallization and crystallization on a seed) with different solvents, a theory of polytypism is expanded (for example, [46]) due to the fact that all modifications known to date have been obtained and structurally characterized for aluminum orthoborates. The largest number of publications is devoted to activated and co-activated $YAl_3(BO_3)_4$ crystals with the space group $R32$. Monoclinic modifications are referred less often, and their centrosymmetry or non-centrosymmetry was not proved but only stated for almost all rare-earth orthoborates. In addition, structural single-crystal studies for the $LnCr_3(BO_3)_4$ having also a variety of modifications are absent to date, a symmetry being affected by the borate:solvent ratios in the batch during the spontaneous crystallization from a flux (Figure 1).

Crystal structures for almost all phases were determined (refined) using the X-ray experiment (there are several works on the neutron experiment, in particular [56,61,66–68]). The occupancies for the *Ln* and *M* sites were not refined (with a few exceptions), they often were fixed to those determined from the ICP elemental analysis. The real composition, obtained by refining the *Ln* and Sc site occupancies in structures of the $LnSc_3(BO_3)_4$ crystals with trigonal and monoclinic symmetries, is generally not coincide and coincide with the charge composition, respectively. As a result, a congruent melting for the monoclinic phases is possible.

Many questions remain to the rare-earth scandium orthoborates, since it is possible to find literature data with the opposite results on existence or absence of different modifications, in particular, a realization of the trigonal phases with the space group *R*32 or *P*321. This is primarily due to the fact that these phases are obtained by different methods under different conditions, and quite possibly, this modification can be obtained by a specific growth method under specific synthesis conditions. We failed to obtain the huntite-family $LnSc_3(BO_3)_4$ compounds with the *Ln* = Sm and Gd by the solid-phase synthesis of the corresponding oxides at the *T* = 1000, 1250, 1500 °C and *T* = 1500 °C, respectively, although in the literature, there is data on the synthesis of the $LnSc_3(BO_3)_4$ compounds with the *Ln* = Sm, Eu, Gd, Y. An X-ray study of single crystals with the nominal composition $LnSc_3(BO_3)_4$ with the *Ln* = La, Ce, Pr, Nd, Tb and numerous solid solutions, grown by the Czochralski method, with the subsequent crystallochemical analysis of the results obtained, makes it possible to doubt the possibility of obtaining the $LnSc_3(BO_3)_4$ with *Ln* = Sm, Gd, Eu, Y by this method. Although internal solid solutions with these components are possible to obtain by a careful selection of the initial charge compositions and synthesis conditions.

When crystals are grown by the Czochralski method from melts at high temperatures, their symmetry depends on a composition and sintering temperature of a charge; a type, composition, symmetry, and orientation of a seed; a growth atmosphere, rotation, and pulling rates, as well as on an efficiently control both processes of B_2O_3 vapors condensation on the growing crystal and temperature gradients above the crucible and in the melt [35]. It should be noted that the $LnSc_3(BO_3)_4$ can easy change its symmetry (group–subgroup), therefore a structural experiment should be performed on single-crystal objects with the analysis of diffraction reflections, including low-intensity ones (the use of high-resolution transmission electron microscopy and synchrotron radiation is also can be carried out), and the refinement of crystallographic site occupancies, i.e., real crystal composition. The refinement of structures by the full-profile method is correct only when using a developed methodology with criteria formulated to attribute the structure to the space group *R*32 or *P*321. This is important for establishing correlations between symmetry, real composition of objects and growth methods and conditions; correct explanation of functional properties observed; direct growth of crystals with a required combination of physical parameters, as well as for clarifying and summarizing crystallochemical data for the huntite family compounds and other functional materials.

6. Patents

Kuz'micheva, G.M.; Podbel'sky, V.V.; Chuykin, N.K.; Kaurova, I.A. Program for the Investigation of the Dynamics of Changes in the Structural Parameters of Compounds with Different Symmetry; Certificate of state registration of computer software no. 2017619941: Moscow, Russia, 12 September 2017 (in Russian).

Author Contributions: Conceptualization, G.M.K.; Methodology, G.M.K. and V.B.R.; Software, V.V.P.; Validation, G.M.K., I.A.K. and V.B.R.; Formal Analysis, G.M.K., I.A.K. and V.B.R.; Investigation, G.M.K. and V.B.R.; Resources, G.M.K., I.A.K., V.B.R. and V.V.P.; Data Curation, G.M.K. and I.A.K.; Writing—Original Draft Preparation, G.M.K. and I.A.K.; Writing—Review & Editing, G.M.K.; Visualization, V.V.P.; Supervision, G.M.K. and I.A.K.; Project Administration, I.A.K.

Funding: This research received no external funding.

Conflicts of Interest: The authors declare no conflict of interest.

References

1. Mills, A.D. Crystallographic data for new rare earth borate compounds, $RX_3(BO_3)_4$. *Inorg. Chem.* **1962**, *1*, 960–961. [CrossRef]
2. Liang, K.C.; Chaudhury, R.P.; Lorenz, B.; Sun, Y.Y.; Bezmaternykh, L.N.; Gudim, I.A.; Temerov, V.L.; Chu, C.W. Magnetoelectricity in the system $RAl_3(BO_3)_4$ (R = Tb, Ho, Er, Tm). *J. Phys. Conf. Ser.* **2012**, *400*, 032046. [CrossRef]
3. Földvári, I.; Beregi, E.; Baraldi, A.; Capelletti, R.; Ryba-Romanowski, W.; Dominiak-Dzik, G.; Munoz, A.; Sosa, R. Growth and spectroscopic properties of rare-earth doped $YAl_3(BO_3)_4$ single crystals. *J. Lumin.* **2003**, *102*, 395–401. [CrossRef]
4. Leonyuk, N.I.; Maltsev, V.V.; Volkova, E.A.; Pilipenko, O.V.; Koporulina, E.V.; Kisel, V.E.; Tolstik, N.A.; Kurilchik, S.V.; Kuleshov, N.V. Crystal growth and laser properties of new $RAl_3(BO_3)_4$ (R = Yb, Er) crystals. *Opt. Mater.* **2007**, *30*, 161–163. [CrossRef]
5. Chen, Y.; Lin, Y.; Gong, X.; Huang, J.; Luo, Z.; Huang, Y. Acousto-optic Q-switched self-frequency-doubling Er:Yb:$YAl_3(BO_3)_4$ laser at 800 nm. *Opt. Lett.* **2012**, *37*, 1565–1567. [CrossRef]
6. Leonyuk, N.I.; Koporulina, E.V.; Barilo, S.N.; Kurnevich, L.A.; Bychkov, G.L. Crystal growth of solid solutions based on the $YAl_3(BO_3)_4$, $NdAl_3(BO_3)_4$ and $GdAl_3(BO_3)_4$ borates. *J. Cryst. Growth* **1998**, *191*, 135–142. [CrossRef]
7. Liang, K.C.; Chaudhury, R.; Lorenz, B.; Sun, Y.; Bezmaternykh, L.; Temerov, V.; Chu, C. Giant magnetoelectric effect in $HoAl_3(BO_3)_4$. *Phys. Rev. B* **2011**, *83*, 180417. [CrossRef]
8. He, J.; Zhang, S.; Zhou, J.; Zhong, J.; Liang, H.; Sun, S.; Huang, Y.; Tao, Y. Luminescence properties of an orange-red phosphor $GdAl_3(BO_3)_4$: Sm^{3+} under VUV excitation and energy transfer from Gd^{3+} to Sm^{3+}. *Opt. Mater.* **2015**, *39*, 81–85. [CrossRef]
9. Li, X.; Wang, Y. Synthesis of $Gd_{1-x}Tb_xAl_3(BO_3)_4$ ($0.05 \leq x \leq 1$) and its luminescence properties under VUV excitation. *J. Lumin.* **2007**, *122*, 1000–1002. [CrossRef]
10. Lokeswara Reddya, G.V.; Rama Moorthya, L.; Chengaiaha, T.; Jamalaiaha, B.C. Multi-color emission tunability and energy transfer studies of $YAl_3(BO_3)_4$:Eu^{3+}/Tb^{3+} phosphors. *Ceram. Int.* **2014**, *40*, 3399–3410. [CrossRef]
11. Yang, F.; Liang, Y.; Liu, M.; Li, X.; Wu, X.; Wang, N. $YAl_3(BO_3)_4$:Tm^{3+}, Dy^{3+}: A potential tunable single-phased white-emitting phosphors. *Opt. Int. J. Light Electron Opt.* **2013**, *124*, 2004–2007. [CrossRef]
12. Brenier, A.; Tu, C.; Zhu, Z.; Wu, B. Red-green-blue generation from a lone dual-wavelength $GdAl_3(BO_3)_4$:Nd^{3+} laser. *Appl. Phys. Lett.* **2004**, *84*, 2034–2036. [CrossRef]
13. Kim, K.; Moon, Y.M.; Choi, S.; Jung, H.K.; Nahm, S. Luminescent properties of a novel green-emitting gallium borate phosphor under vacuum ultraviolet excitation. *Mater. Lett.* **2008**, *62*, 3925–3927. [CrossRef]
14. Volkov, N.V.; Gudim, I.A.; Eremin, E.V.; Begunov, A.I.; Demidov, A.A.; Boldyrev, K.N. Magnetization, magnetoelectric polarization, and specific heat of $HoGa_3(BO_3)_4$. *JETP Lett.* **2014**, *99*, 67–75. [CrossRef]
15. Popova, E.; Leonyuk, N.; Popova, M.; Chukalina, E.; Boldyrev, K.; Tristan, N.; Klingeler, R.; Büchner, B. Thermodynamic and optical properties of $NdCr_3(BO_3)_4$. *Phys. Rev. B* **2007**, *76*, 054446. [CrossRef]
16. Szytuła, A.; Przewoźnik, J.; Żukrowski, J.; Prokhorov, A.; Chernush, L.; Zubov, E.; Dyakonov, V.; Duraj, R.; Tyvanchuk, Y. On the peculiar properties of triangular-chain $EuCr3 (BO3) 4$ antiferromagnet. *J. Solid State Chem.* **2014**, *210*, 30–35.
17. Boldyrev, K.N.; Chukalina, E.P.; Leonyuk, N.I. Spectroscopic investigation of rare-earth chromium borates $RCr_3(BO_3)_4$ (R = Nd, Sm). *Phys. Solid State* **2008**, *50*, 1681–1683. [CrossRef]
18. Kadomtseva, A.M.; Popov, Y.F.; Vorob'ev, G.P.; Pyatakov, A.P.; Krotov, S.S.; Kamilov, K.I.; Ivanov, V.Y.; Mukhin, A.A.; Zvezdin, A.K.; Kuz'menko, A.M.; et al. Magnetoelectric and magnetoelastic properties of rare-earth ferroborates. *Low Temp. Phys.* **2010**, *36*, 511–521. [CrossRef]
19. Mukhin, A.A.; Vorob'ev, G.P.; Ivanov, V.Y.; Kadomtseva, A.M.; Narizhnaya, A.S.; Kuz'menko, A.M.; Popov, Y.F.; Bezmaternykh, L.N.; Gudim, I.A. Colossal magnetodielectric effect in $SmFe_3(BO_3)_4$ multiferroic. *JETP Lett.* **2011**, *93*, 275–281. [CrossRef]
20. Zvezdin, A.K.; Vorob'ev, G.P.; Kadomtseva, A.M.; Popov, Y.F.; Pyatakov, A.P.; Bezmaternykh, L.N.; Kuvardin, A.V.; Popova, E.A. Magnetoelectric and magnetoelastic interactions in $NdFe_3(BO_3)_4$ multiferroics. *JETP Lett.* **2006**, *83*, 509–514. [CrossRef]

21. Wang, G.F. Structure, growth, nonlinear optics, and laser properties of $RX_3(BO_3)_4$ (R = Y, Gd, La; X = Al, Sc). In *Structure-Property Relationships in Non-Linear Optical Crystals I. Structure and Bonding*; Wu, X.T., Chen, L., Eds.; Springer: Berlin, Germany, 2012; Volume 144, pp. 105–120.

22. Huber, G. Solid-State Laser Materials. In *Laser sources and Applications*; Miller, A., Finlayson, D.M., Eds.; Institute of Physics: Bristol, UK, 1996; pp. 141–162.

23. Noginov, M.A.; Noginova, N.E.; Caulfield, H.J.; Venkateswarlu, P.; Thompson, T.; Mahdi, M.; Ostroumov, V. Short-pulsed stimulated emission in the powders of $NdAl_3(BO_3)_4$, $NdSc_3(BO_3)_4$, and $Nd:Sr_5(PO_4)_3F$ laser crystals. *J. Opt. Soc. Am. B Opt. Phys.* **1996**, *13*, 2024–2033. [CrossRef]

24. Sardar, D.K.; Castano, F.; French, J.A.; Gruber, J.B.; Reynolds, T.A.; Alekel, T.; Keszler, D.A.; Clark, B.L. Spectroscopic and laser properties of Nd^{3+} in $LaSc_3(BO_3)_4$ host. *J. Appl. Phys.* **2001**, *90*, 4997–5001. [CrossRef]

25. Li, Y.; Aka, G.; Kahn-Harari, A.; Vivien, D. Phase transition, growth, and optical properties of $Nd_xLa_{1-x}Sc_3(BO_3)_4$ crystals. *J. Mater. Res.* **2001**, *16*, 38–44. [CrossRef]

26. Dobretsova, E.A.; Boldyrev, K.N.; Borovikova, E.Y.; Chernyshev, V.A. Structural and optical properties of $Nd_xGd_{1-x}Cr_3(BO_3)_4$ solid solutions. *EPJ Web Conf.* **2017**, *132*, 03012. [CrossRef]

27. Leonyuk, N.I.; Leonyuk, L.I. Growth and characterization of $RM_3(BO_3)_4$ crystals. *Prog. Cryst. Growth Charact. Mater.* **1995**, *31*, 179–278. [CrossRef]

28. Leonyuk, N.I. Half a century of progress in crystal growth of multifunctional borates $RAl_3(BO_3)_4$ (R = Y, Pr, Sm–Lu). *J. Cryst. Growth* **2017**, *476*, 69–77. [CrossRef]

29. Leonyuk, N.I. Growth of new optical crystals from boron-containing fluxed melts. *Crystallogr. Rep.* **2008**, *53*, 511–518. [CrossRef]

30. Wang, G.; Lin, Z.; Hu, Z.; Han, T.P.J.; Gallagher, H.G.; Wells, J.R. Crystal growth and optical assessment of $Nd^{3+}:GdAl_3(BO_3)_4$ crystal. *J. Cryst. Growth* **2001**, *233*, 755–760. [CrossRef]

31. Ye, N.; Stone-Sundberg, J.L.; Hruschka, M.A.; Aka, G.; Kong, W.; Keszler, D.A. Nonlinear Optical Crystal $Y_xLa_ySc_z(BO_3)_4$ (x + y + z = 4). *Chem. Mater.* **2005**, *17*, 2687–2692. [CrossRef]

32. Liu, H.; Li, J.; Fang, S.H.; Wang, J.Y.; Ye, N. Growth of $YAl_3(BO_3)_4$ crystals with tungstate based flux. *Mater. Res. Innov.* **2011**, *15*, 102–106. [CrossRef]

33. Wang, G.; Han, T.P.J.; Gallagher, H.G.; Henderson, B. Crystal growth and optical properties of $Ti^{3+}:YAl_3(BO_3)_4$ and $Ti^{3+}:GdAl_3(BO_3)_4$. *J. Cryst. Growth* **1997**, *181*, 48–54. [CrossRef]

34. Beregi, E.; Watterich, A.; Madarász, J.; Tóth, M.; Polgár, K. X-ray diffraction and FTIR spectroscopy of heat treated $R_2O_3:3Ga_2O_3:4B_2O_3$ systems. *J. Cryst. Growth* **2002**, *237*, 874–878. [CrossRef]

35. Durmanov, S.T.; Kuzmin, O.V.; Kuzmicheva, G.M.; Kutovoi, S.A.; Martynov, A.A.; Nesynov, E.K.; Nesynov, E.K.; Panyutin, V.L.; Rudnitsky, Y.P.; Smirnov, G.V.; et al. Binary rare-earth scandium borates for diode-pumped lasers. *Opt. Mater.* **2001**, *18*, 243–284. [CrossRef]

36. Wang, G.; Han, T.P.J.; Gallagher, H.G.; Henderson, B. Novel laser gain media based on Cr^{3+}-doped mixed borates $RX_3(BO_3)_4$. *Appl. Phys. Lett.* **1995**, *67*, 3906–3908. [CrossRef]

37. Wang, G.; Gallagher, H.G.; Han, T.P.J.; Henderson, B. The growth and optical assessment of Cr^{3+}-doped $RX(BO_3)_4$ crystals with R = Y, Gd; X = Al, Sc. *J. Cryst. Growth* **1996**, *163*, 272–278. [CrossRef]

38. Kutovoi, S.A.; Laptev, V.V.; Matsnev, S.Y. Lanthanum scandoborate as a new highly efficient active medium of solid-state lasers. *Quantum Electron.* **1991**, *21*, 131–132. [CrossRef]

39. Kutovoi, S.A. Growth and Laser Properties of Lanthanum-Scandium Borate Single Crystals with Rare-Earth Activators. Ph.D. Thesis, Moscow State Academy of Fine Chemical Technology, Moscow, Russia, 1998. (In Russian)

40. Wang, Y.H.; Li, X.X. Synthesis and photoluminescence properties of $LnAl_3(BO_3)_4:Eu^{3+}$ (Ln = La^{3+}, Gd^{3+}) under UV and VUV excitation. *J. Electrochem. Soc.* **2006**, *153*, G238–G241. [CrossRef]

41. Plachinda, P.A.; Belokoneva, E.L. High temperature synthesis and crystal structure of new representatives of the huntite family. *Cryst. Res. Technol.* **2008**, *43*, 157–165. [CrossRef]

42. Dobretsova, E.A.; Borovikova, E.Y.; Boldyrev, K.N.; Kurazhkovskaya, V.S.; Leonyuk, N.I. IR spectroscopy of rare-earth aluminum borates $RAl_3(BO_3)_4$ (R = Y, Pr-Yb). *Opt. Spectrosc.* **2014**, *116*, 77–83. [CrossRef]

43. Hong, H.P.; Dwight, K. Crystal structure and fluorescence lifetime of $NdAl_3(BO_3)_4$, a promising laser material. *Mater. Res. Bull.* **1974**, *9*, 1661–1665. [CrossRef]

44. Wang, G.; Meiyun, H.; Luo, Z. Structure of β-$NdAl_3(BO_3)_4$ (NAB) crystal. *Mater. Res. Bull.* **1991**, *26*, 1085–1089. [CrossRef]

45. Kurazhkovskaya, V.S.; Borovikova, E.Y.; Leonyuk, N.I.; Koporulina, E.V.; Belokoneva, E.L. Infrared spectroscopy and the structure of polytypic modifications of $RM_3(BO_3)_4$ borates (R—Nd, Gd, Y; M—Al, Ga, Cr, Fe). *J. Struct. Chem.* **2008**, *49*, 1035–1041. [CrossRef]

46. Belokoneva, E.L.; Timchenko, T.I. Polytype relationships in borate structures with the general formula $YAl_3(BO_3)_4$, $NdAl_3(BO_3)_4$, and $GdAl_3(BO_3)_4$. *Sov. Phys. Crystallogr.* **1983**, *28*, 658–661.

47. Belokoneva, E.L. The structures of new germanates, gallates, borates, and silicates with laser, piezoelectric, ferroelectric, and ion-conducting properties. *Russ. Chem. Rev.* **1994**, *63*, 533–549. [CrossRef]

48. Jarchow, O.; Lutz, F.; Klaska, K.H. Polymophie and Fehlordnung von $NdAl_3(BO_3)_4$. *Z. F. Krist* **1979**, *149*, 162.

49. Lutz, F.; Huber, G. Phosphate and borate crystals for high optical gain. *J. Cryst. Growth* **1981**, *52*, 646–649. [CrossRef]

50. Ballman, A.A. New series of synthetic borates isostructural with carbonate mineral huntite. *Am. Mineral.* **1962**, *47*, 1380.

51. Dobretsova, E.A.; Boldyrev, K.N.; Chernyshev, V.A.; Petrov, V.P.; Mal'tsev, V.V.; Leonyuk, N.I. Infrared spectroscopy of europium borates $EuM_3(BO_3)_4$ (M = Al, Cr, Fe, Ga) with a huntite mineral type of structure. *Bull. Rus. Acad. Sci. Phys.* **2017**, *81*, 546–550. [CrossRef]

52. Yang, F.G.; Zhu, Z.J.; You, Z.Y.; Wang, Y.; Li, J.F.; Sun, C.L.; Cao, J.F.; Ji, Y.X.; Wang, Y.Q.; Tu, C.Y. The growth, thermal and nonlinear optical properties of single-crystal $GdAl_3(BO_3)_4$. *Laser Phys.* **2011**, *21*, 750–754. [CrossRef]

53. Hinatsu, Y.; Doi, Y.; Ito, K.; Wakeshima, M.; Alemi, A. Magnetic and calorimetric studies on rare-earth iron borates $LnFe_3(BO_3)_4$ (Ln = Y, La–Nd, Sm–Ho). *J. Solid State Chem.* **2003**, *172*, 438–445. [CrossRef]

54. Joubert, J.C.; White, W.B.; Roy, R. Synthesis and crystallographic data of some rare earth–iron borates. *J. Appl. Crystallogr.* **1968**, *1*, 318–319. [CrossRef]

55. Campa, J.A.; Cascales, C.; Gutierrez-Puebla, E.; Monge, M.A.; Rasines, I.; Ruiz-Valero, C. Crystal structure, magnetic order, and vibrational behavior in iron rare-earth borates. *Chem. Mater.* **1997**, *9*, 237–240. [CrossRef]

56. Ritter, C.; Vorotynov, A.; Pankrats, A.; Petrakovskii, G.; Temerov, V.; Gudim, I.; Szymczak, R. Magnetic structure in iron borates $RFe_3(BO_3)_4$ (R = Er, Pr): A neutron diffraction and magnetization study. *J. Phys. Condens. Matter* **2010**, *22*, 206002. [CrossRef] [PubMed]

57. Popova, M.N. Optical spectroscopy of low-dimensional rare-earth iron borates. *J. Magn. Magn. Mater.* **2009**, *321*, 716–719. [CrossRef]

58. Popova, M.N. Spectroscopy of compounds from the family of rare-earth orthoborates. *J. Rare Earths* **2009**, *27*, 607–611. [CrossRef]

59. Fischer, P.; Pomjakushin, V.; Sheptyakov, D.; Keller, L.; Janoschek, M.; Roessli, B.; Schefer, J.; Petrakovskii, G.; Bezmaternikh, L.; Temerov, V.; et al. Simultaneous antiferromagnetic Fe^{3+} and Nd^{3+} ordering in $NdFe_3(^{11}BO_3)_4$. *J. Phys. Condens. Matter* **2006**, *18*, 7975. [CrossRef]

60. Fausti, D.; Nugroho, A.A.; van Loosdrecht, P.H.; Klimin, S.A.; Popova, M.N.; Bezmaternykh, L.N. Raman scattering from phonons and magnons in $RFe_3(BO_3)_4$. *Phys. Rev. B* **2006**, *74*, 024403. [CrossRef]

61. Ritter, C.; Pankrats, A.; Gudim, I.; Vorotynov, A. Determination of the magnetic structure of $SmFe_3(BO_3)_4$ by neutron diffraction: Comparison with other $RFe_3(BO_3)_4$ iron borates. *J. Phys. Condens. Matter* **2012**, *24*, 386002. [CrossRef]

62. Popova, M.N.; Malkin, B.Z.; Boldyrev, K.N.; Stanislavchuk, T.N.; Erofeev, D.A.; Temerov, V.L.; Gudim, I.A. Evidence for a collinear easy-plane magnetic structure of multiferroic $EuFe_3(BO_3)_4$: Spectroscopic and theoretical studies. *Phys. Rev. B* **2016**, *94*, 184418. [CrossRef]

63. Klimin, S.A.; Fausti, D.; Meetsma, A.; Bezmaternykh, L.N.; Van Loosdrecht, P.H.M.; Palstra, T.T.M. Evidence for differentiation in the iron-helicoidal chain in $GdFe_3(BO_3)_4$. *Acta Crystallogr. Sect. B Struct. Sci.* **2005**, *61*, 481–485. [CrossRef]

64. Levitin, R.Z.; Popova, E.A.; Chtsherbov, R.M.; Vasiliev, A.N.; Popova, M.N.; Chukalina, E.P.; Klimin, S.A.; van Loosdrecht, P.H.M.; Fausti, D.; Bezmaternykh, L.N. Cascade of phase transition in $GdFe_3(BO_3)_4$. *J. Exp. Theor. Phys. Lett.* **2004**, *79*, 423–426. [CrossRef]

65. Klimin, S.A.; Kuzmenko, A.B.; Kashchenko, M.A.; Popova, M.N. Infrared study of lattice dynamics and spin-phonon and electron-phonon interactions in multiferroic $TbFe_3(BO_3)_4$ and $GdFe_3(BO_3)_4$. *Phys. Rev. B* **2016**, *93*, 054304. [CrossRef]

66. Ritter, C.; Balaev, A.; Vorotynov, A.; Petrakovskii, G.; Velikanov, D.; Temerov, V.; Gudim, I. Magnetic structure, magnetic interactions and metamagnetism in terbium iron borate TbFe$_3$(BO$_3$)$_4$: A neutron diffraction and magnetization study. *J. Phys. Condens. Matter* **2007**, *19*, 196227. [CrossRef]

67. Ritter, C.; Pankrats, A.; Gudim, I.; Vorotynov, A. Magnetic structure of iron borate DyFe$_3$(BO$_3$)$_4$: A neutron diffraction study. *J. Phys. Conf. Ser.* **2012**, *340*, 012065. [CrossRef]

68. Ritter, C.; Vorotynov, A.; Pankrats, A.; Petrakovskii, G.; Temerov, V.; Gudim, I.; Szymczak, R. Magnetic structure in iron borates RFe$_3$(BO$_3$)$_4$ (R = Y, Ho): A neutron diffraction and magnetization study. *J. Phys. Condens. Matter* **2008**, *20*, 365209. [CrossRef]

69. Malakhovskii, A.V.; Sokolov, V.V.; Sukhachev, A.L.; Aleksandrovsky, A.S.; Gudim, I.A.; Molokeev, M.S. Spectroscopic properties and structure of the ErFe$_3$(BO$_3$)$_4$ single crystal. *Phys. Solid State* **2014**, *56*, 2056–2063. [CrossRef]

70. Borovikova, E.Y.; Dobretsova, E.A.; Boldyrev, K.N.; Kurazhkovskaya, V.S.; Maltsev, V.V.; Leonyuk, N.I. Vibrational spectra and factor group analysis of rare-earth chromium borates, RCr$_3$(BO$_3$)$_4$, with R = La–Ho. *Vib. Spectrosc.* **2013**, *68*, 82–90. [CrossRef]

71. Kurazhkovskaya, V.S.; Dobretsova, E.A.; Borovikova, E.Y.; Mal'tsev, V.V.; Leonyuk, N.I. Infrared spectroscopy and the structure of rare-earth chromium borates RCr$_3$(BO$_3$)$_4$ (R = La-Er). *J. Struct. Chem.* **2011**, *52*, 699. [CrossRef]

72. Dobretsova, E.A.; Boldyrev, K.N.; Popova, M.N.; Gavrilkin, S.Y.; Mukhin, A.A.; Ivanov, V.Y.; Mal'tsev, V.V.; Leonyuk, N.I.; Malkin, B.Z. Phase transitions and exchange interactions in the SmCr$_3$(BO$_3$)$_4$ crystal. *EPJ Web Conf.* **2017**, *132*, 02008. [CrossRef]

73. Borovikova, E.Y.; Boldyrev, K.N.; Aksenov, S.M.; Dobretsova, E.A.; Kurazhkovskaya, V.S.; Leonyuk, N.I.; Savon, A.E.; Deyneko, D.V.; Ksenofontov, D.A. Crystal growth, structure, infrared spectroscopy, and luminescent properties of rare-earth gallium borates RGa$_3$(BO$_3$)$_4$, R = Nd, Sm–Er, Y. *Opt. Mater.* **2015**, *49*, 304–311. [CrossRef]

74. Parthe, E.; Hu, S.Z. β-LaSc$_3$(BO$_3$)$_4$: Correction of the crystal structure. *Mater. Res. Innov.* **2003**, *7*, 353–354. [CrossRef]

75. He, M.; Wang, G.; Lin, Z.; Chen, W.; Lu, S.; Wu, Q. Structure of medium temperature phase β-LaSc$_3$(BO$_3$)$_4$ crystal. *Mater. Res. Innov.* **1999**, *2*, 345–348. [CrossRef]

76. Goryhnov, A.V.; Kuz'micheva, G.M.; Mukhin, B.V.; Zharikov, E.V.; Ageev, A.Y.; Kutovoj, S.A.; Kuz'min, O.V. An X-ray diffraction study of LaSc$_3$(BO$_3$)$_4$ crystals activated with chromium and neodymium ions. *Russ. J. Inorg. Chem.* **1996**, *41*, 1531–1536.

77. Lebedev, V.A.; Pisarenko, V.F.; Chuev, Y.M.; Zhorin, V.V.; Perfilin, A.A.; Shestakov, A.V. Synthesis and study of non-linear laser crystals CeSc$_3$(BO$_3$)$_4$. *Adv. Solid State Lasers* **1996**. [CrossRef]

78. Fedorova, M.V.; Kononova, N.G.; Kokh, A.E.; Shevchenko, V.S. Growth of MBO$_3$ (M = La, Y, Sc) and LaSc$_3$(BO$_3$)$_4$ crystals from LiBO$_2$-LiF fluxes. *Inorg. Mater.* **2013**, *49*, 482–486. [CrossRef]

79. Wang, G.; He, M.; Chen, W.; Lin, Z.; Lu, S.; Wu, Q. Structure of low temperature phase γ-LaSc$_3$(BO$_3$)$_4$ crystal. *Mater. Res. Innov.* **1999**, *2*, 341–344. [CrossRef]

80. Magunov, I.R.; Efryushina, N.P.; Voevudskaya, S.V.; Zhikhareva, E.A.; Zhirnova, A.P. Preparation and properties of double borates of scandium and REE of the cerium subgroup. *Inorg. Mater.* **1986**, *21*, 1337–1341.

81. Peterson, G.A.; Keszler, D.A.; Reynolds, T.A. Stoichiometric, trigonal huntite borate CeSc$_3$(BO$_3$)$_4$. *Int. J. Inorg. Mater.* **2000**, *2*, 101–106. [CrossRef]

82. Kuz'micheva, G.M.; Kaurova, I.A.; Rybakov, V.B.; Podbel'sky, V.V.; Chuykin, N.K. Structural Instability in Single-Crystal Rare-Earth Scandium Borates RESc$_3$(BO$_3$)$_4$. *Cryst. Growth Des.* **2018**, *18*, 1571–1580. [CrossRef]

83. Kuz'micheva, G.M.; Rybakov, V.B.; Kutovoi, S.A.; Kuz'min, O.V.; Panyutin, V.L. Morphotropic Series of LnSc$_3$(BO$_3$)$_4$ Compounds. *Crystallogr. Rep.* **2000**, *45*, 910–915. [CrossRef]

84. Rybakov, V.B.; Kuz'micheva, G.M.; Mukhin, B.V.; Zharikov, E.V.; Ageev, A.Yu.; Kutovoi, S.A.; Kuz'min, O.V. X-ray diffraction study of huntite-family (Ce,Gd)Sc$_3$(BO$_3$)$_4$ compounds. *Russ. J. Inorg. Chem.* **1997**, *42*, 9–16.

85. Kuz'min, O.V.; Kuz'micheva, G.M.; Kutovoi, S.A.; Martynov, A.A.; Panyutin, V.L.; Chizhikov, V.I. Cerium scandium borate—An active nonlinear medium for diode-pumped lasers. *Quantum Electron.* **1998**, *28*, 50. [CrossRef]

86. Kuz'micheva, G.M.; Rybakov, V.B.; Novikov, S.G.; Ageev, A.Yu.; Kutovoi, S.A.; Kuz'min, O.V. Disordered structures of rare-earth scandium borates of the huntite family. *Russ. J. Inorg. Chem.* **1999**, *44*, 352–366.

87. Kuz'micheva, G.M.; Kutovoi, S.A.; Rybakov, V.B.; Kuz'min, O.V.; Panyutin, V.L. X-ray diffraction method for the determination of solid-solution compositions of rare-earth scandium borates belonging to the huntite family. *Russ. J. Inorg. Chem.* **2005**, *50*, 1160–1168.

88. Reynolds, T.A. Synthetic, Structural, and Spectroscopic Investigations of Acentric Laser Hosts and Ionic Optical Converters. Ph.D. Thesis, Oregon State University, Corvallis, OR, USA, 1992.

89. Rybakov, V.B.; Kuzmicheva, G.M.; Zharikov, E.V.; Ageev, A.Y.; Kutovoi, S.A.; Kuz'min, O.V. Crystal structure of NdSc$_3$(BO$_3$)$_4$. *Russ. J. Inorg. Chem.* **1997**, *41*, 1594–1601.

90. Sváb, E.; Beregi, E.; Fábián, M.; Mészáros, G. Neutron diffraction structure study of Er and Yb doped YAl$_3$(BO$_3$)$_4$. *Opt. Mater.* **2012**, *34*, 1473–1476. [CrossRef]

91. Kuzmicheva, G.M. *Some Aspects of the Applied Crystallochemistry*; MIREA: Moscow, Russia, 2016. (In Russian)

92. Farrugia, L.J. WinGX suite for small-molecule single-crystal crystallography. *J. Appl. Crystallogr.* **1999**, *32*, 837–838. [CrossRef]

93. Sheldrick, G.M. *SHELXT*—Integrated space-group and crystal-structure determination. *Acta Crystallogr.* **2015**, *A71*, 3–8. [CrossRef] [PubMed]

94. Kuz'micheva, G.M.; Podbel'sky, V.V.; Chuykin, N.K.; Kaurova, I.A. *Program for the Investigation of the Dynamics of Changes in the Structural Parameters of Compounds with Different Symmetry*; Certificate of State Registration of Computer Software No. 2017619941: Moscow, Russia, 2017. (In Russian)

95. Brandenburg, K. *Diamond*; Crystal Impact GbR: Bonn, Germany, 1999.

96. Shannon, R.D. Revised effective ionic radii and systematic studies of interatomic distances in halides and chalcogenides. *Acta Crystallogr. Sect. A* **1976**, *32*, 751–767. [CrossRef]

97. Beregi, E.; Sajó, I.; Lengyel, K.; Bombicz, P.; Czugler, M.; Földvári, I. Polytypic modifications in heavily Tb and Eu doped gadolinium aluminum borate crystals. *J. Cryst. Growth* **2012**, *351*, 72–76. [CrossRef]

98. Kuz'micheva, G.M. Stability of coordination polyhedra and mechanisms of stabilization of structural types (for compounds with rare-earth elements). In *Problems of Crystallochemistry*; Nauka: Moscow, Russia, 1989; pp. 15–47. (In Russian)

99. Kuz'micheva, G.M.; Voloshin, A.E.; Eliseev, A.A. On the stability of coordination polyhedra and structural types in chalcogenides of rare-earth elements. *Russ. J. Inorg. Chem.* **1984**, *29*, 1374–1378. (In Russian)

100. Meyn, J.P.; Jensen, T.; Huber, G. Spectroscopic properties and efficient diode-pumped laser operation of neodymium-doped lanthanum scandium borate. *IEEE J. Quantum Electron.* **1994**, *30*, 913–917. [CrossRef]

101. Xu, X.; Ye, N. Gd$_x$La$_{1-x}$Sc$_3$(BO$_3$)$_4$: A new nonlinear optical crystal. *J. Cryst. Growth* **2011**, *324*, 304–308. [CrossRef]

102. Ye, N. Structure design and crystal growth of UV nonlinear borate materials. In *Structure-Property Relationships in Non-Linear Optical Crystals I*; Springer: Berlin, Germany, 2012; pp. 181–221.

103. Gheorghe, L.; Khaled, F.; Achim, A.; Voicu, F.; Loiseau, P.; Aka, G. Czochralski growth and characterization of incongruent melting La$_x$Gd$_y$Sc$_z$(BO$_3$)$_4$ (x + y + z = 4) nonlinear optical crystal. *Cryst. Growth Des.* **2016**, *16*, 3473–3479. [CrossRef]

104. Gheorghe, L.; Achim, A.; Voicu, F. A new promising nonlinear optical (NLO) crystal for visible and ultraviolet (UV) regions. *AIP Conf. Proc.* **2012**, *1472*, 141–147.

105. Hruschka, M.A. A New Trigonal Huntite Material and Subgroup Relationships between Crystallographic Space Groups. Ph.D. Thesis, Oregon State University, Corvallis, OR, USA, 2005.

106. Ye, N.; Zhang, Y.; Chen, W.; Keszler, D.A.; Aka, G. Growth of nonlinear optical crystal Y$_{0.57}$La$_{0.72}$Sc$_{2.71}$(BO$_3$)$_4$. *J. Cryst. Growth* **2006**, *292*, 464–467. [CrossRef]

107. Keszler, D.A.; Stone-Sundberg, J.L.; Ye, N.; Hruschka, M.A. *U.S. Patent No. 7,534,377 B2*; U.S. Patent and Trademark Office: Washington, DC, USA, 19 May 2009.

108. Stone-Sundberg, J.L. A Contribution to the Development of Wide Band-Gap Nonlinear Optical and Laser Materials. Ph.D. Thesis, Oregon State University, Corvallis, OR, USA, 2001.

109. Bourezzou, M.; Maillard, A.; Maillard, R.; Villeval, P.; Aka, G.; Lejay, J.; Loiseau, P.; Rytz, D. Crystal defects revealed by Schlieren photography and chemical etching in nonlinear single crystal LYSB. *Opt. Mater.Express* **2011**, *1*, 1569–1576. [CrossRef]

110. Li, W.; Huang, L.; Zhang, G.; Ye, N. Growth and characterization of nonlinear optical crystal Lu$_{0.66}$La$_{0.95}$Sc$_{2.39}$(BO$_3$)$_4$. *J. Cryst. Growth* **2007**, *307*, 405–409. [CrossRef]

111. Xu, X.; Wang, S.; Ye, N. A new nonlinear optical crystal Bi$_x$La$_y$Sc$_z$(BO$_3$)$_4$ (x + y + z = 4). *J. Alloys Compd.* **2009**, *481*, 664–667. [CrossRef]

112. Maczka, M.; Pietraszko, A.; Hanuza, J.; Majchrowski, A. Raman and IR spectra of noncentrosymmetric $Bi_{0.21}La_{0.91}Sc_{2.88}(BO_3)_4$ single crystal with the huntite-type structure. *J. Raman Spectrosc.* **2010**, *41*, 1297–1301. [CrossRef]

113. Kokh, A.E.; Kuznetsov, A.B.; Pestryakov, E.V.; Maillard, A.; Maillard, R.; Jobard, C.; Kononova, N.G.; Shevchenko, V.S.; Kragzhda, A.A.; Uralbekov, B.; et al. Growth of the complex borates $Y_xR_ySc_{2+z}(BO_3)_4$ (R = Nd, Pr, x + y + z = 2) with huntite structure. *Cryst. Res. Technol.* **2017**, *52*, 1600371. [CrossRef]

114. Kaurova, I.A.; Gorshkov, D.M.; Kuz'micheva, G.M.; Rybakov, V.B. Composition, structure and symmetry of huntite-family compounds. *Fine Chem. Technol.* **2018**, *13*, 58–67. (In Russian)

115. Nakajima, H.; Usui, T.; Joly, Y.; Suzuki, M.; Wakabayashi, Y.; Kimura, T.; Tanaka, Y. Quadrupole moments in chiral material $DyFe_3(BO_3)_4$ observed by resonant x-ray diffraction. *Phys. Rev. B* **2016**, *93*, 144116. [CrossRef]

116. Usui, T.; Tanaka, Y.; Nakajima, H.; Taguchi, M.; Chainani, A.; Oura, M.; Shin, S.; Katayama, N.; Sawa, H.; Wakabayashi, Y.; et al. Observation of quadrupole helix chirality and its domain structure in $DyFe_3(BO_3)_4$. *Nat. Mater.* **2014**, *13*, 611. [CrossRef] [PubMed]

117. Bubnova, R.S.; Filatov, S.K. *High-Temperature Crystal Chemistry of Borates and Borosilicates*; Nauka: St. Petersburg, Russia, 2008. (In Russian)

118. Chernov, A.A. Growth of copolymer chains and mixed crystals: Statistics of trials and errors. *Usp. Fiz. Nauk* **1970**, *13*, 111–162. (In Russian) [CrossRef]

119. Shtukenberg, A.G.; Punin, Y.O.; Frank-Kamenetskaya, O.V. The kinetic ordering and growth dissymmetrisation in crystalline solid solutions. *Russ. Chem. Rev.* **2006**, *75*, 1083–1106. [CrossRef]

120. Kuz'micheva, G.M.; Kaurova, I.A.; Zagorul'ko, E.A.; Bolotina, N.B.; Rybakov, V.B.; Brykovskiy, A.A.; Zharikov, E.V.; Lis, D.A.; Subbotin, K.A. Structural perfection of $(Na_{0.5}Gd_{0.5})MoO_4$:Yb laser crystals. *Acta Mater.* **2015**, *87*, 25–33. [CrossRef]

121. Volkov, V.; Rico, M.; Mandez-Blas, A.; Zaldo, C. Preparation and Properties of Disordered $NaBi(XO_4)_2$ X = W or Mo Crystals Doped with Rare Earths. *J. Phys. Chem. Solids* **2002**, *63*, 95–105. [CrossRef]

122. Cascales, C.; Serrano, M.D.; Esteban-Betegón, F.; Zaldo, C.; Peters, R.; Petermann, K.; Huber, G.; Ackermann, L.; Rytz, D.; Dupre, C.; et al. Structural, spectroscopic, and tunable laser properties of Yb^{3+}-doped $NaGd(WO_4)_2$. *Phys. Rev. B* **2006**, *74*, 174114. [CrossRef]

123. Belokoneva, E.L.; Leonyuk, N.I.; Pashkova, A.V.; Timchenko, T.I. New modifications of rare earth-aluminium borates. *Kristallografiya* **1988**, *33*, 1287–1288. (In Russian)

124. Verma, A.; Krishna, P. *Polymorphism and Polytypism in Crystals*; Mir: Moscow, USSR, 1969. (In Russian)

125. Dornberger-Schiff, K. Grundzuge einer theorie von OD-Strukturen aus Schichten. *Abh. Deutsch. Akad. Wiss. Berl.* **1964**, *3*, 1–107.

MDPI

St. Alban-Anlage 66

4052 Basel

Switzerland

Tel. +41 61 683 77 34

Fax +41 61 302 89 18

www.mdpi.com

Crystals Editorial Office

E-mail: crystals@mdpi.com

www.mdpi.com/journal/crystals

www.ingramcontent.com/pod-product-compliance
Lightning Source LLC
Chambersburg PA
CBHW051914210326
41597CB00033B/6137